THE ENERGY MANAGERS' HANDBOOK

THE ENERGY MANAGERS' HANDBOOK

Second edition

Gordon A. Payne
BSc, CEng, FInstE

Westbury House

Published by Westbury House, the books division of IPC Science and Technology Press Limited, PO Box 63, Bury Street, Guildford, Surrey GU2 5BH, England

First published 1977
Reprinted 1979
Second edition 1980

ISBN 0 86103 032 X (cloth)
ISBN 0 86103 033 8 (paper)
(ISBN 0 902852 73 6 1st edition, cloth)
(ISBN 0 902852 59 0 1st edition, paper)

British Library Cataloguing in Publication Data
Payne, Gordon A
The energy managers' handbook. – 2nd ed.
I. Energy conservation – Great Britain – Management
658.2'6 TJ163.4.G7

ISBN 0-86103-032-X
ISBN 0-86103-033-8 Pbk

Printed in Great Britain by
Thomson Litho Ltd, East Kilbride, Scotland

CONTENTS

FOREWORD TO SECOND EDITION

British industry is responding to the challenges and opportunities created by high energy prices.

A constantly growing number of companies can prove that investment in energy conservation equipment and efficient energy management bring swift and satisfactory savings in energy and cash. And there is the creation of a virtually brand new industrial sector engaged in manufacturing and marketing energy-saving equipment, technology and services.

But many more firms could cut costs and increase their competitiveness and profits if they used energy more efficiently.

A large part of the savings achieved by industry have been due to the efforts of energy managers.

I am particularly pleased, therefore, to note the publication of this new, expanded version of *The Energy Managers' Handbook*. It can be expected, like the first edition, to serve all those concerned with energy management as a useful guide and comprehensive source of reference.

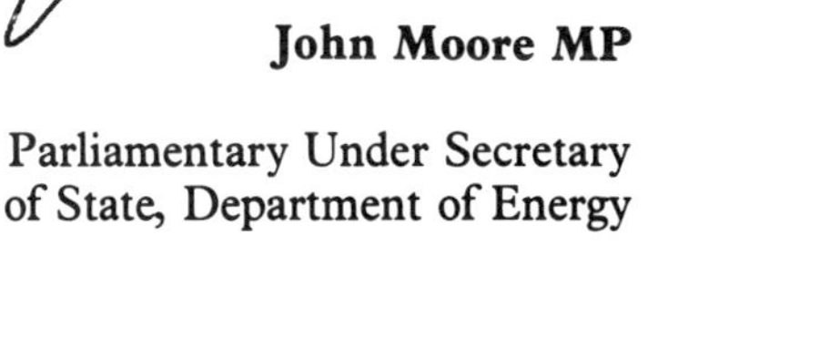

John Moore MP

Parliamentary Under Secretary of State, Department of Energy

INTRODUCTION

The need to conserve our declining fuel resources, especially of oil and natural gas, is now widely accepted. By moderating our demand for energy we can usefully extend the period of time available for the safe and proper development of alternative sources. But for many sectors of industry and commerce there is a more immediate need to conserve energy in order to contain escalating costs and maintain financial viability. Large manufacturing organisations like steelworks and glassworks have long kept a watchful eye on their use of energy, but every organisation, no matter how small, needs now to follow suit.

The handbook is written for the many individuals responsible, perhaps for the first time, for the management of energy and who may not always have previous experience in this field. Unless their approach is systematic and comprehensive, much of their effort is likely to be wasted. The handbook is designed to provide, in a single publication, sufficient guidance on the principles involved, supported by check lists and illustrations, to enable readers to generate and sustain a straightforward energy conservation programme tailored to their own particular needs.

The handbook is not intended to be a text book of fuel technology and it does not deal with any topic in greater depth than is necessary to establish the basic principles, and perhaps whet the appetite. Fortunately the technology and techniques required to carry through a programme are generally well known and documented. Many sources of information were consulted in preparing the manuscript and may thanks are due to the authors not only of established books but also of the many booklets, articles and reprints on energy which have been published in recent years.

Illustrations used in the handbook are individually acknowledged. Many equipment manufacturers have been most helpful in providing photographs and other material and I am grateful not only to them but also to other friends in the equipment industry who let me have data of all sorts but whose products are not illustrated.

In compiling a handbook of this sort I have been very aware of the debt that I owe to a number of former colleagues and close friends who, over many years, helped to shape my views and who overlooked my inadequacies. To those who, for my benefit, explained complex things in simple language I am particularly grateful.

Finally my thanks are due to my publisher for their encouragement when, with some trepidation, I first approached them with the idea of yet another - but I sincerely hope, different - book on energy, and for their cooperation since.

My wife, Jean, assisted and supported me throughout, typing the manuscript several times and providing other invaluable secretarial help, all of which I acknowledge with gratitude. Our joint efforts will, we hope, in some way help to preserve some of the precious resources which we now enjoy.

Guildford, Surrey
April 1976

Gordon Payne

INTRODUCTION TO THE SECOND EDITION

Although the general format is much the same, the opportunity has been taken in this second edition slightly to expand the content of the handbook, to include several topics not previously covered and to update some text and illustrations.

I am grateful to those who kindly wrote to me following publication of the first edition, to offer encouragement and to point out shortcomings, which, I trust, I have now eradicated.

Uffculme, Devon
February 1980

Gordon Payne

1 ENERGY CONSERVATION

1.1 The need to conserve

As we begin to appreciate that energy is vital to man's survival, it is not surprising that the topic is of increasing interest. There are many differing views on energy resources, future requirements and costs, but several basic ideas on energy are now firmly established.

As reserves of fossil fuels (mainly coal, oil and natural gas) are depleted, huge amounts of new capital will be required for further and more difficult exploration and production.

As the world population increases and as underdeveloped regions of the world rightly strive for better living standards, the total demand for energy will continue to rise substantially.

Sufficient energy may not be available to meet our long-term demands especially if developments in nuclear energy fall behind schedule or if supplies of more conventional fuels are disrupted for political or other reasons.

New discoveries of energy may provide local security of supply for a period, but as world prices increase they are unlikely to be cheap sources for very long.

Over half of the energy used by man is wasted. There is ample scope for improvement, though there are often practical difficulties and the more drastic energy-saving measures might require unacceptable changes to our present life style. Improving efficiency and making fossil fuels last longer will give more time for the proper and safe development of nuclear power and of renewable sources of energy like solar, wind and tidal power, though it may be years before these begin to make a significant contribution to the world's energy supply, and some may never prove to be economic on a large scale.

This handbook is not primarily concerned, however, with forecasts of demand and other broad economic matters, important though these are, nor with advanced new technologies. It assumes that there *is* a real need to conserve conventional forms of energy, especially in industry and commerce, and it deals with some of the practical aspects that must be considered. Industry and commerce should require little convincing of the need to conserve in order to cut their costs and to remain competitive. A comparison of their present energy prices with those of just a few years ago should suffice to demonstrate the rapid rise that has already occurred and, with it, a shift in emphasis towards operating costs as distinct from the capital cost of energy-using plant.

Industry and commerce account for an important part of the world's total consumption of energy, although, of course, effective conservation in these sectors depends also on action by many others, including governments, local authorities, major utilities and transport undertakings, as well as private individuals. But conservation by industry and commerce is important not only because they have an immediate financial incentive, but also because they are in a unique position as users, suppliers and employers to influence and encourage others.

When many costs are escalating and difficult to control, energy is often one area where effective savings can be made and costs can be contained. The comforting thing for the Energy Manager, or whoever is responsible for energy within an organisation, is that a great deal can be done by applying well-established principles and techniques. No white-hot new technology is required. What *is* required is the dedicated and systematic application of proven ideas.

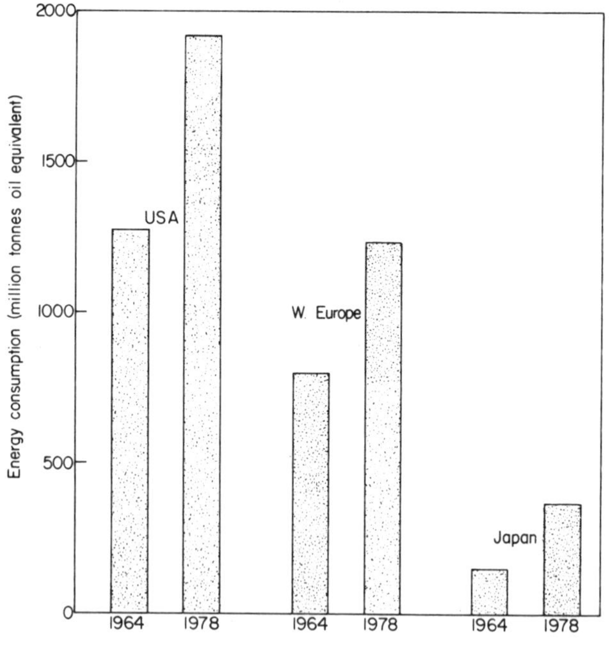

1/1 Increase in the total consumption of energy for some major industrial areas, 1964–78.
Based on data from the *BP statistical review of the world oil industry.*

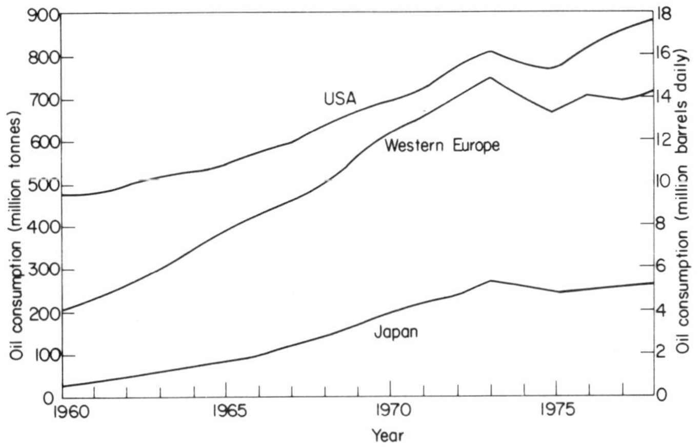

1/2 Oil consumption 1960–1978. An indication of the need for energy conservation, and, from 1973, the effect on oil consumption of the level of world industrial activity. Based on data from the *BP statistical review of the world oil industry*.

1.2 Principles of energy conservation

Certain overriding principles apply to all forms of energy conservation and an appreciation of these is important at the outset.

The manner and extent of *all* energy use should be challenged, including the appropriateness of the process method and the size of the plant involved. Incidental benefits and penalties should be carefully evaluated.

If possible, useful work should be done at each stage of temperature or pressure reduction. Most energy is lost ultimately to the outside environment in the form of heat and it should be made to do as much work as possible on the way.

Energy saving is greatly assisted by measurement. Consistent units and definitions should be used if measurements and comparisons are to be meaningful.

Heat which is recovered must be usable and an end-use must be found *before* recovery is seriously contemplated. In a normal commercial situation the value of the saving must clearly exceed the cost of recovery.

Apparent energy cost savings should be closely examined to ensure that they have not resulted in increased costs elsewhere. Savings must be *real* ones.

Waste in all forms not only squanders human effort, time and materials, it also uses energy. Reduction in waste is especially desirable where the materials involved have high intrinsic energy contents. Metals, glass, plastics, paper and refractories are examples of high-energy materials. Improvements in design which prolong the lifespan of such materials make a valuable contribution to energy conservation.

Undetected recycling during manufacture wastes energy, especially recycling of materials (e.g. laundry, die-castings) that can easily be reprocessed.

1.3 An energy-saving programme

For effective energy conservation, attention needs to be focused on energy use, but this should not be considered in isolation. Regard must be paid to other important aspects as an overall programme is developed and the following items should be included.

Energy supply security.
Safety and health considerations.
Capital required or capital saved.
Attitudes of employees at all levels.
Environmental effects of any changes.
Maintenance of plant and equipment.
Emergency or standby requirements.

The *real* purpose of each piece of energy-using plant should be assessed before any change is made. A low thermal efficiency as well as a low return on investment may be quite acceptable, for example, on a rarely used standby generator if it provides 'insurance' against shut-down or serious interruption to production. A sophisticated piece of automatic plant may operate at high thermal efficiency, but it may become a serious liability if skilled labour is not available to maintain it, or if it produces some environmental or safety hazard. The use of a single grade of fuel of the lowest quality suitable for the process, bought from a single supplier and stored in the minimum quantity may seem attractive at first sight. But such an arrangement could result in reduced purchasing power, a risk of interruption in supply and possible environmental problems. These aspects need careful evaluation.

1.4 The main areas for examination

In seeking energy savings there are a number of possible areas to be considered and the following are used in later sections of this handbook.

Management of energy.
Delivery, storage and handling of fuels.

Boilers.
Steam and hot water utilisation.
Industrial space heating.
Furnaces and driers.
Flue gases and heat recovery.
Electricity.
Services.
Road transport.

In each section the main text deals with the principles involved and provides background facts and figures. It is followed by a summarising checklist to assist the reader to identify energy-saving opportunities in his (or her) own organisation and to help him to develop a cohesive energy conservation programme.

1.5 Energy measurement

Without the measurement of energy input and use, full control is impossible, yet many organisations spend large sums of money on fuel and little on

1/3 A compact portable time recorder being used to indicate periods of operation of a heating plant. The same instrument can be used to monitor several sites, but similar instruments can be permanently mounted, if desired.
Servis Recorders (Industrial) Ltd, Gloucester.

instruments to measure and control. Measurements should be reliable and accurate, but no more accurate than necessary to do the job and, in the initial stages of a programme, even approximate values are often useful. The basic equipment needed is fairly straightforward, and there is merit in simplicity and robustness.

Permanently installed instruments should be used for all larger plant on which frequent or continuous measurements are required. They may be used to indicate or record, or to control. These instruments should be regarded as an integral part of the plant and should be of good quality. They should be carefully installed to protect them from dirt and damage and to facilitate maintenance. The principal measurements made are usually of temperatures, pressures and flow rates and the oxygen and carbon dioxide contents of waste gases.

For each instrument giving visual readings it is essential first to consider the following in order that a correct choice can be made.

Why is it needed?
How will the readings be used?
What accuracy is necessary?
Who will read it?
Where can it be suitably located?
What will be the conditions of use?

Portable test equipment may be used to take measurements where such fixed instrumentation cannot be justified, for situations where permanent instruments would be vulnerable to damage and also for *ad hoc* investigations and tests. Numbers of 'spot' measurements to predetermined schedules can more readily be taken if pipes and ducts are permanently fitted with suitable test points.

Although accuracy is sometimes sacrificed, a good range of portable test instruments is available and a selection is described in Appendix B. Portable instruments should be robust and should preferably be used and maintained by one person only.

For every measurement it is essential that the method used should ensure that it is truly representative, otherwise the result will be worthless. Gas sampling requires particular care if erroneous results are not to be obtained.

1.6 Thermal efficiency

The term 'thermal efficiency' should generally be used in its strict scientific sense; that is, the percentage of energy supplied which is usefully used in any instance. It should be related to the plant as a whole and should not be confused with 'quality of combustion'. To refer to the efficiency of a burner alone, no matter how perfect its combustion performance, is meaningless — its thermal

1/4 Checks on the thermal efficiency of combustion plant can be carried out relatively easily using portable instruments to measure the oxygen (or carbon dioxide) content of the flue gases and their temperature and smoke content. Here, the oxygen content of flue gases is being measured with an electronic indicator.
Telegan Ltd, Croydon CR9 2BU.

efficiency is *nil* until the heat that it produces is utilised in some way.

For boilers and similar plant, thermal efficiency can conveniently be stated as the percentage of heat input that is transferred to the steam, water or other working fluid. For furnaces, kilns, etc. it is often more practical to relate product output to energy input; that is, to state the overall specific energy consumption (e.g. kWh/ton melted, MJ/1000 bricks).

The conditions under which thermal efficiency is determined must always be clearly stated or the standard or test code quoted. In particular, the heating value of the fuel should always be defined as the gross or net calorific value (CV) at constant volume or pressure. The gross or upper value is that obtained by cooling the products of combustion to a defined ambient temperature and condensing any water vapour that they contain. The net or lower calorific value excludes the heat obtained by condensing the water vapour. Thus combustion plant appears more efficient when net CV values are used in calculations rather than gross CV values.

Care is needed in comparing the performance of plant fired with fuels of differing hydrogen content and therefore differing water vapour content in their waste gases. Care must also be exercised in using efficiency values for particular items of plant based on full-load tests under ideal conditions. The average practical efficiency of a steam boiler, for example, may be much lower than its tested efficiency when losses from blowdown and variable-load operation are taken into account.

Efficiency of heat utilisation is as important as efficiency of heat production, and overall efficiency, which is what *really* matters, is the product of the separate efficiencies in each step of the heat production, distribution and utilisation chain. It follows that if a step can be eliminated or made more efficient the overall efficiency will be improved proportionately.

A selection of typical calorific values is given in Appendix A.

2 MANAGEMENT OF ENERGY

2.1 Management and employees

As costs rise and as the supply and use of energy in industry and commerce require more careful long-term planning, senior management must increasingly concern themselves with the subject. No organisation can expect to maintain its position if it fails to ensure that its future supply of energy is secure and that there is close control of its utilisation.

Senior management should demonstrate a firm commitment to energy conservation from the outset and the following basic facts should be established.

The approximate total amount of each form of energy used. The figures should cover the various grades of coal and oil, natural gas, LPG (butane and propane) and bought-in electricity. The quantities of fuel used should be compared by converting them all to a common unit of energy, such as GJ, kWh or therms.

The present average net cost of each form of energy compared with earlier years, again using a common energy unit, such as pence/GJ.

A first estimate of the savings which might be possible by taking immediate steps to effect improvements without capital expenditure.

The approximate relative importance of capital and operating costs for energy-using plant over recent years. Increasing operating costs will clearly justify greater future spending on more efficient plant.

Real interest in cutting energy costs needs to be generated throughout the organisation at an early stage. This may be done by means of items in company journals and bulletins, posters and stickers, training programmes, films, suggestion schemes and so on. Results achieved must also be communicated and proper recognition given to employees' ideas and contributions. Employees at all levels should know that savings are necessary to offset soaring energy costs and to maintain the organisation's viability rather than to increase profits. They will be painfully aware of the increasing costs of heating their own homes and running their cars, so that giving them practical guidance on these things is an obvious way to stimulate personal interest.

Energy conservation affects employees. It may add to their workloads, require their cooperation, affect their general comfort or even their pay packets. But

2/1 All employees should be aware of the need to cut energy costs. Interest can be stimulated by using lively, bold posters and stickers. These posters are available free. *Department of Energy, London SW1P 4QJ.*

---

energy conservation will also improve the viability of the organisation and with it job security and pay prospects. Health and safety requirements must, of course, still be met in any changes to improve efficiency. However, because energy conservation focuses attention on systematic measurement and control, health and safety will often be improved.

Energy conservation also has important public relations aspects. Most companies try to publicise their achievements locally, if not nationally. Energy-saving case histories can provide good material for press releases and the opportunity should not be overlooked to engender a little public goodwill and perhaps counteract other outside pressures to which businesses are sometimes subjected.

The detailed work required to improve energy efficiency, set targets, monitor progress, keep abreast of technical developments and so on should be delegated. Line managers and supervisors should have clear direction on their responsibilities within their own departments but overall coordination and direction should be vested in one person — the Energy Manager. In a small organisation these duties may be undertaken part-time or the Energy Manager may simply be the person nominated to liaise with a qualified outside consultant, retained to develop an energy management programme and periodically to review performance.

2.2 The Energy Manager

In a larger organisation, it is desirable, though not essential, that the Energy Manager (or whatever his or her title may be) should be a qualified fuel technologist or engineer with some experience of costing methods and the financial evaluation of capital projects. But principally the Energy Manager needs a questioning mind and the ability to command the support of colleagues. Terms of reference should be clearly defined, particularly whether or not the Energy Manager has any authority within service or production departments or whether he acts only in an advisory and coordinating role. He should report direct to a senior executive and should have free access to information on energy matters. His precise duties will depend upon the size and type of organisation in which he works but they should include some or all of the following:

To generate interest in energy conservation and to sustain this interest with new ideas and activities. To lecture to training courses and to outside organisations and to give short practical talks on various technical topics.

To maintain essential summaries of energy purchases, stocks and consumption. Regularly to review energy utilisation performance and trends and to advise senior management.

To be the focal point for departmental records of energy use and to ensure that the records and accounting systems are uniform and in consistent units.

To coordinate the efforts of all energy users and to set tough but realistic targets. To give technical advice on energy-saving equipment and techniques, or to identify suitable sources of sound technical guidance.

To identify where major energy waste is occurring, to quantify the losses in financial terms and to put forward practical recommendations for reducing them.

To identify areas of activity which require detailed study and to concentrate efforts on these. To maintain a record of all in-depth studies and to review progress.

To provide a basic manual or handbook of good energy practice, using published material or, if this is unsuitable, to produce material specifically to meet the needs of his own organisation.

To give specialist advice to purchasing, planning, production and other functions on the longer-term aspects of energy conservation.

To ensure that, in making improvements or implementing suggestions from any source, health and safety are not jeopardised.

To liaise with committees and working groups within his own industry and, with his company's approval, to exchange ideas on cost-cutting techniques and performance figures in similar processes.

To maintain contact with appropriate research organisations, equipment manufacturers and professional bodies to ensure that he remains up-to-date on all significant developments in the field of energy conservation.

To remain up-to-date on world and national energy developments and to advise senior management on energy matters generally.

The initial workload may be heavy and consideration should be given to using outside consultancy services. This should not be regarded as a reflection on the Energy Manager's ability but rather a recognition that an outside specialist will probably have useful ideas to contribute and will be able to help develop an energy management programme more quickly than might otherwise be possible. For smaller organisations, particularly, a consultant retained to review progress *regularly* and to comment critically 'from the outside' will also help to support the Energy Manager, build up his expertise over a period and confirm to him that he is working on the right lines.

2.3 Energy auditing

Information on existing energy purchases, distribution and use is essential before any improvements or future planning can be seriously contemplated. The form of data collection will vary but certain features will be common to all organisations.

The purpose of an audit is to establish the basic relative costs of various forms of energy, their main uses and the principal points at which there is waste or inefficiency. Detailed studies and energy balances of specific items of plant or systems should be undertaken quite separately and should not be permitted to delay an initial audit.

An initial audit need not be very sophisticated or accurate. The aim should be to obtain results quickly and to improve accuracy in the light of experience. Analysis forms will probably have to be redesigned several times.

The first collection of data will almost certainly reveal inadequacies in metering and measurement. These should be noted, and steps taken to make improvements.

Obvious examples of waste (leaks, lack of insulation, unnecessary operations, etc.) will be found in the course of even the most elementary check. These should be attended to *at once.* A few quick results will be a great encouragement to further action and will demonstrate the value of the work to the sceptical.

Checks should be carried out at night, at weekends, and during holidays as well as during normal daytime working to ensure that nothing is overlooked.

The time period over which audits take place should be clearly defined so that meaningful comparisons can be made.

Particular attention should be given to identifying (and if possible quantifying) peaks in energy consumption as these usually attract high tariff charges or result in additional plant requirements and overheads.

An initial audit should concentrate on energy *inputs* to departments and individual items of plant in order to find out where and how much energy is used. However, energy *outputs* will also be important in some cases (e.g. boilers), and, starting with the major items, heat balances should then be drawn up for these.

Comparisons can best be made if a common unit of heat content is used (e.g. GJ, kWh or therms). Some conversion factors are given in Appendix A, and others may be found in various standard reference books. Because of the low thermal efficiency of most public electricity generation, about 10.5 MJ (10 000 Btu) of heat energy is required to produce 1 kWh of electricity, but the heat equivalent of 1 kWh is only 3.6 MJ (3412 Btu). The higher values certainly represent the energy cost to society as a whole, but for an organisation assessing potential savings in monetary terms the lower values should be used.

The precise form of data collection will depend upon the size of the organisation, its functions and its complexity, but most energy audits can conveniently be undertaken in several phases.

A. Fuel and energy purchases.
B. Energy used by various departments.
C. Energy used in various processes within departments.
D. Energy waste checklists (in conjunction with C).

The headings used in the examples of analysis forms (pp 14-17) are for guidance only and users should develop forms to suit their *own* particular needs.

The limitations of energy auditing should be recognised. It may tell us a great deal about our use of energy, but it is of little practical value if we do not *also*

A. Energy purchases summary

Period from to Reference

(1)	(2)	(3)	(4)	(5)	(6)	(7)	(8)	(9)
Energy	*Grade or specification*	*Unit*	*Total purchased*	*Total cost*	*Average cost per unit (5) ÷ (4)*	*Calorific value (net)*	*Total energy content (4) x (7)*	*Average cost per standard energy unit (5) ÷ (8)*
Coal								
Oil								
Gas								
LPG								
Electricity								
Other								
Prepared by		Overall cost (10)			Overall energy use (11)			
Date		Notes						
Examined by								
Date								

- A form of this sort should be used to make periodic overall energy reviews.
- The data provides a general indication only of an organisation's energy use and costs, as a starting point for conservation.
- It takes no account of handling and storage costs, convenience, premium values, etc, or of the efficiency of use.
- For cost comparisons between different fuels, total heat contents (8) should be in the same standard units. Use the conversions in Appendix A.
- Overall energy use (11) can also more readily be monitored if standard units are used.

B. Energy used by departments

Period from   to   Reference

(1)	(2)		(3)	(4)	(5)	(6)	(7)	(8)	(9)	(10)	(11)
Department	*Section*	*Energy*	*Coal*	*Oil*	*Gas*	*LPG*	*Elec.*	*Steam*	*Total standard units ( ) (3) to (8)*	*Output or production index*	*Energy productivity (9) ÷ (10)*
		Grade									
		Unit (M)									
		M to S	X	X	X	X	X	X			
		M									
		S									
		M									
		S									
		M									
		S									
		M									
		S									
		M									
		S									
		M									
		S									
		M									
		S									
		M									
		S									
(12) Overall measured units		Σ M									
(13) Overall standard units		Σ S									
Prepared by		Notes									
Date											
Examined by											
Date											

- Conversion from measured units (M) to standard energy units (S), e.g. kWh or Btu, greatly facilitates comparisons. Include conversion factors (M to S) on the printed form for each energy source and state the standard unit under (9).
- Energy productivity (11) should be monitored only if production or output is significantly related to energy use.
- Columns (3) to (8) should be designed to suit the organisation's fuel pattern.
- Overall energy use in standard units (13) should be related to overall production, weather conditions, etc.

C. Energy end-use survey

Department

Reference

Survey by

Period of survey from hr on until hr on

(1)	(2)	(3)	(4)		(5)		(6)	(7)	(8)
Plant or end-use	*Ref. or code*	*Energy type*	*Energy consumption rate*				*Peak month*	*Thermal efficiency (estimate E)*	*Comments*
			Average	h/day	*Maximum*	*Time*			

- It is important to record the precise period during which the survey is carried out.
- Note should be made of known daily consumption peaks, especially of electricity. The maximum consumption rates and times at which these occur should be recorded.
- Known seasonal peaks should also be noted.
- The survey will help to identify major energy-using plant, assist comprehensive listing, and indicate shortcomings in measurement and control.
- Estimate efficiency if no figures are available and follow up with actual tests. 'E' in column.
- Note any obvious shortcomings and record on form D.

D. Energy waste checklist

Department

Reference

(1)	(2)	(3)	(4)	(5)	(6)	(7)
Plant or end-use	*Ref. or code*	*Waste noted and action required*	*Noted by*	*Date*	*Action by*	*Date*

- All identified energy waste should be recorded and a follow-up made to ensure that action has been taken.
- Specific waste noted during general surveys should be entered on form D for action, otherwise it may be overlooked.

know what equipment is available and what techniques to apply in order to make improvements.

2.4 Energy costs

Although the conservation of energy is a desirable objective in its own right, most industrial and commercial users will expect improvements in efficiency to show savings in hard cash. The true cost of various forms of energy at each point of use is required if the precise cost-effectiveness of each piece of plant is to be assessed and if future planning is to be reliable. Accounting for every heat unit from purchase until it is finally discarded can be a mammoth task requiring detailed analysis. In practice, various short-cuts are employed and cost figures for various forms of energy can be derived by successive approximations until reasonably reliable values are obtained. An approximate cost figure or inspired guess is better than nothing!

A significant part of the total cost of energy is its actual purchased net cost. Energy is usually treated as other materials, requirements being specified by engineering or production departments and purchased by a specialist department at the best terms that can be negotiated. In practice, mains gas and electricity are often supplied on fixed tariffs with, at first sight, little scope for negotiation. Oil, liquefied petroleum gases and sometimes solid fuels are more usually bought in a competitive market. Energy purchasing can be fairly complex and it needs to be coordinated.

There is no univeral 'best buy' as far as energy is concerned. A continuing audit of energy purchases will enable prices to be compared on a common basis but every form of energy has particular advantages and disadvantages to the user and *he alone* can assess their value in each case.

Once broad energy policy has been determined, however, certain things can be done to reduce costs. Technical considerations often preclude such changes, but close scrutiny of tariffs and price scales and detailed discussions with suppliers may reveal opportunities to lower costs by such things as:

Changing to a fuel or grade in more plentiful supply.
Providing a more uniform offtake.
Accepting larger deliveries.
Accepting off-peak or seasonal deliveries.
Providing improved delivery facilities.
Speeding up delivery procedures and reducing demurrage.
Negotiating interruptible supply arrangements.
Changing to a more advantageous tariff.
Rationalising the number of fuel grades used.
Negotiating group or bigger bulk supplies.

The buyer should consider periodically, in as detached a way as he can, the real value of his business to his various suppliers. He should consider *their* problems with seasonal fluctuations in demand, transport, labour, competition, etc., and then seek to make his business more attractive in return for cash advantages. The buyer should know each supplier's most pressing problem and main weakness, as well as likely long-term trends in energy availability and price.

In addition to the purchased net cost, the true cost of each form of energy should take account of all items directly connected with fuel or energy storage, distribution, conversion and utilisation and should include such things as:

Storage tanks or fuel bunkers.
Land for storage.
Energy-using plant and firing equipment.
Flues, chimneys and dampers.
Pumps, heaters, fans and other ancillaries.
Water for steam and for cooling purposes.
Water treatment and effluent treatment plant.
Electricity for ancillary equipment.
Wages and salaries.
Maintenance and cleaning.

The identification and costing of these items will provide a better appreciation of where the fixed and variable costs associated with energy use lie and will facilitate the evaluation of true energy costs as a proportion of total costs.

2.5 Planning

Operating costs are now more significant than they were for energy-using plant. Combustion performance, optimum plant sizing and load planning to give high overall efficiency are all important aspects to be considered when planning production schedules or installing new plant. Planning now requires a much more searching look at energy use, and certain questions need to be answered in the early stages before detailed evaluation and return on investment calculations are started.

Is the process method being considered the most efficient and economic bearing in mind likely future energy costs? Would a change in heat transfer method (e.g. from radiation to forced convection) be beneficial?

Would complete plant redesign at higher cost yield greater savings than modification of existing arrangements?

What is the precise purpose of the plant? Could alternative lower-energy methods (e.g. mechanical removal of water instead of drying by heat) produce the same results?

Is the plant of optimum size to ensure that it is always fully loaded? Is other process plant matched to it?

Is the *operation* of plant scheduled to minimise electrical and other peaks?

What use can be made of waste heat? Will new plans result in additional waste heat and if so, can this be recovered?

Can any process step be eliminated or combined with another?

Each organisation will have its own criteria for evaluating capital projects and determining the return on investment. These should be applied in the normal way, but only after careful consideration of likely energy requirements of the various alternatives. The calculation should then be based on considered estimates of likely *future* energy costs over the expected life of the plant.

In the longer term, increasing interest will need to be taken in the overall energy requirements of finished products and services, according to carefully defined raw material and energy inputs and processing routes. As energy costs rise, the value of low-energy products with high durability will increase. Organisations that provide products or services with high energy requirements need to be especially aware of likely cost trends and alternatives. Energy analysis of this sort has been carried out in a number of industries.

2.6 Regular control and monitoring

Once the more urgent tasks of eliminating waste by simple 'good housekeeping' methods have been completed, following an initial energy audit, regular reviews of performance should continue — energy auditing is *not* a one-off exercise.

Data from these successive audits should be used to make comparisons of performance — of one period with another, one department with another, one building with another and so on. Care is needed. Other conditions as well as energy use must be reasonably comparable or suitable corrections and allowances must be made.

Comparisons are facilitated if energy consumption is related to one or more appropriate parameters, for example GJ/tonne output, kWh/1000 items to give specific energy consumption or energy productivity. When a single product is manufactured or simple service provided this is fairly easy, but mixed activities make such comparisons difficult. Selection of the most suitable parameter to which to relate energy consumption deserves more consideration than it usually gets. After correcting for weather conditions, should we, for example, relate energy used for space heating to building volume, total floor area or the number of people in the building? The choice may be difficult, but at least we should be consistent. Comparisons of this sort are particularly useful in organisations with large numbers of very similar premises or activities.

Overall figures can be misleading. Variations in output may also result in variations in specific energy consumption, in much the same way as boiler efficiency changes with load. A greatly improved performance in one department may also balance, and hide, a deterioration elsewhere. Energy records are best prepared by individual departments and related to their own output if this is a significant factor. In this way improvements and lapses can be detected and, more importantly, the *reasons* established more readily.

In the control of energy costs, attention should be given to the overheads associated with the use of energy as well as to the actual amount and cost of energy saved. We should also clearly distinguish between fixed and variable costs. When improved efficiency results in lower energy consumption, actual cash savings will generally be less than proportionate *unless* the additional effective plant capacity created by the improvement is fully utilised. The aim must be always to spread the 'energy overheads' as well as to cut consumption.

For day-to-day control of buildings, plant and processes, as distinct from periodic reviews and comparisons of performance, a schedule of measurements should be compiled. In some instances automatic measurement and control can be justified. For simple manual measurements, suitable test equipment is described and illustrated in Appendix B.

Data collected in this way, on a routine basis, should not be treated in a casual manner but carefully analysed by a competent person who is aware of safety, health and environmental implications, and who is able to make the correct interpretations.

3 DELIVERY, STORAGE AND HANDLING OF FUELS

3.1 Sources of loss

Adequate stocks of fuel, available to meet any emergency, interruption of supply or unexpected peak in demand are valuable assets. But stocks also cost money to maintain and are potential sources of loss if they are not carefully controlled. In addition to physical losses, storage and handling systems have an initial capital cost, require maintenance and supervision, may require heating and occupy space. All these factors must be considered in efforts to minimise overall costs without losing security of supply.

Losses and waste are liable to occur from the moment fuels are delivered and attention to their delivery, storage and handling should form an essential part of every energy conservation programme. Detailed advice on good practice is readily available from most major suppliers who may also have instructional films or other aids. Various national codes and standards contain detailed guidance on good practice. As well as reducing costs, sound procedures also improve safety and aid good combustion.

After delivery, losses can occur in several ways.

Spillage and leakage

Correct measurement of the contents of bunkers and tanks is essential to prevent overfilling. Fortunately, most losses of solid fuel and oil can readily be detected provided regular inspection is undertaken. Action can then be taken before they become serious. Gas leaks are much less obvious but potentially more dangerous. There should be no leaks whatsoever from gas storage and handling systems.

Deterioration

This applies mainly to coal in exposed stockpiles, but the loss of volatile components, oxidation, sludge formation and so on should be minimised as far as possible for *all* fuels.

Pilfering and vandalism

The more valuable the fuel, the greater the risk. Storage areas may need to be securely fenced and controlled. Isolated above-ground storage tanks for distillate oils (e.g. kerosine) are particularly vulnerable. Drain valves should be locked and no unconnected spurs or dead-legs should be permitted in handling systems. Small LPG cylinders should be strictly controlled to prevent losses or illicit swopping of empty for filled cylinders. The unauthorised filling of LPG cylinders from bulk storage without using the correct equipment is extremely dangerous.

In most cases it is unnecessary to check physically all deliveries, but an occasional random check is prudent to ensure that the correct grade of fuel is being delivered and that the full quantity is being discharged. Such a check will take only a few minutes but will discourage casual delivery and acceptance practices. Where detailed quality checks are necessary, it is important that the sampling and test methods used should be to agreed and recognised standards. Duplicate samples should be retained in case disputes arise.

- **Ensure that attention to delivery, storage and handling arrangements forms an important part of any energy conservation programme.**
- **Ensure that storage facilities cannot be overfilled and that regular checks are made for spills and leaks.**
- **Check that fuels are not being stolen. Consider how someone would attempt to pilfer and make things difficult for them.**
- **Check deliveries from time to time to ensure that quality and quantity are correct.**
- **Assess storage and handling costs and allocate these to the appropriate fuels.**

3.2 Oil fuels

Oil delivery lines should if possible be short and self-draining so that, for residual (heated) grades of oil, trace heating is not required. When delivery line heating *is* necessary, it should be thermostatically controlled and applied just before a delivery is due. Traced lines should always be insulated.

When more than one storage tank is available and total storage capacity is ample, deliveries of residual grades of oil fuel should be made into the tank currently in use, the reserve tank(s) being allowed to cool. If this is done, stock control will need to be tighter and the tanks should be changed over periodically. The temperature of oil in reserve should not be allowed to approach its pour point (at which it begins to solidify) and, when required for use, its temperature should be raised slowly — at not more than 1°C (2°F) an hour, but at only one-half or one-third of this rate, if possible. The formation of sludge can be reduced by minimising thermal cycling and by regularly draining water from each tank. This should be done just before a delivery is made and the drain valve should be securely locked afterwards.

Residual grades of oil fuel are heated in storage in order to keep them sufficiently fluid to flow under gravity to the tank outlet where their temperature is raised to the handling temperature for distribution to burners. Tank thermostats should be of good quality and carefully positioned and set. Tank heaters should cover as much of the tank base as practicable and they must *always* be positioned below the lowest oil offtake point. Holding uninsulated tanks at handling temperature — or even worse, at atomising temperature — instead of at a suitable storage temperature is particularly wasteful, and should be avoided by the use of outflow and line heaters.

The correct storage temperature for any particular oil fuel will depend on its low-temperature flow characteristics or 'pumpability'. Advice should be sought from the oil supplier on the precise temperatures to be used. Some standards and codes of practice give guidance; for example, for the UK, BS 2869, '*Petroleum Fuels for Oil Engines and Burners*', recommends the following minimum storage and handling temperatures:

		Storage	*Handling*
Class E	Light fuel oil	10°C (50°F)	10°C (50°F)
Class F	Medium fuel oil	25°C (77°F)	30°C (86°F)
Class G	Heavy fuel oil	40°C (104°F)	50°C (122°F)
Class H	Extra heavy fuel oil	45°C (113°F)	55°C (131°F)

Although specified as minima, these temperatures will in most cases be more than adequate and they should not be exceeded without good reason. It should be appreciated that recommendations are often intended to assist the *design* of storage and handling facilities and to ensure adequate heater capacity to cope with fuels of the worst low-temperature flow characteristics (high pour point fuels) as much as to give guidance on operating temperatures for normal refinery output.

Heat losses from storage tanks depend on a number of factors like size, shape, situation and so on. In addition, because new tanks quickly acquire an inner skin of gelled oil which itself acts as insulation, actual heat losses are generally somewhat lower than calculated values. Estimates can only be approximate and several methods of calculation in current use imply an accuracy far higher than it is possible to obtain in practice. Even an approximate figure will suffice, however, to determine whether or not a tank should be insulated.

Approximate heat losses for above-ground, uninsulated, heated oil storage tanks in reasonably sheltered situations are given by

$H = kA(T - t)$ where

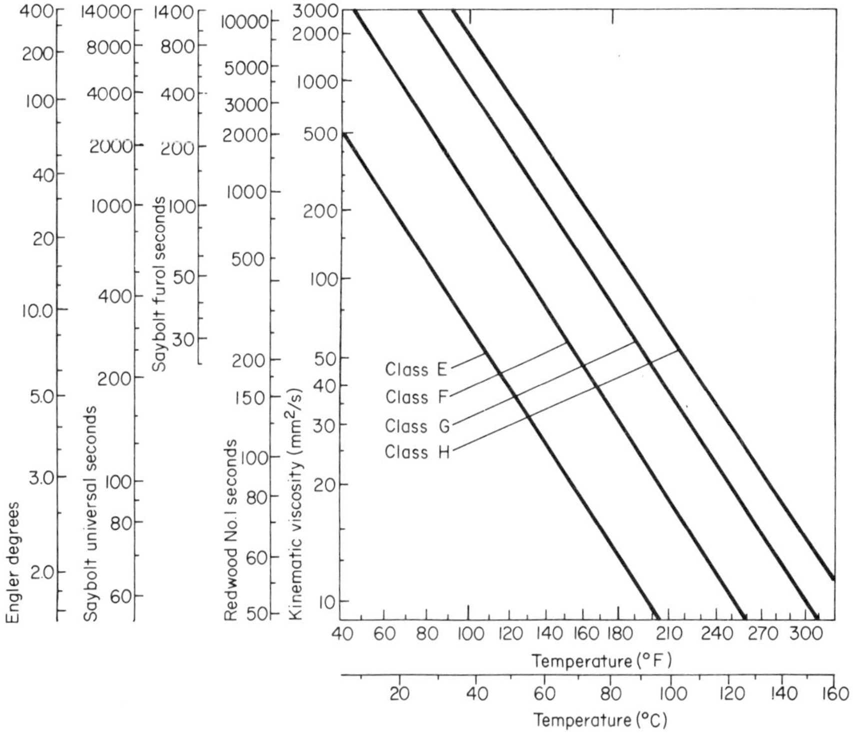

3/1 Viscosity/temperature chart for residual petroleum oil fuels.
British Standards Institution, London W1.

The classes of fuel shown are those specified in BS 2869.

Class E.	Light fuel oil	Class G.	Heavy fuel oil.
Class F.	Medium fuel oil.	Class H.	Extra heavy fuel oil.

Correct viscosity is important for the proper atomisation of oil fuels and the appropriate atomising temperature should be determined from:

(1) The viscosity/temperature characteristics of the oil actually being used, obtained, if necessary, by drawing a line through some fixed point indicated in the specification parallel to those shown, and
(2) The atomising viscosity recommended by the burner manufacturer. When this cannot be obtained, the following are suggested:

Pressure-jet, air and steam atomisers:
20 cSt (20 mm²/s), 90 sec Redwood 1, 3 degrees Engler.

Rotary-cup atomisers:
90 cSt (90 mm²/s), 370 sec Redwood 1, 12 degrees Engler

H =	Average total heat loss rate	Watts	Btu/h
k =	Constant	8.0	1.4
A =	Tank area exposed to air	m^2	ft^2
T =	Oil storage temperature	°C	°F
t =	Average annual ambient temperature	°C	°F

In comparing the cost of insulation with the value of heat saved, the *real* cost of steam and electricity should be used together with the *average* annual air temperature for the locality (in the UK, 10°C, 50°F, is often assumed), *not* the lowest winter temperature. Heat losses can readily be reduced by 80%. Tank insulation should be non-combustible and weatherproof.

As indicated earlier, the handling temperature depends on the grade of fuel being considered. Oil should not be raised to atomising temperature on leaving the storage tank unless the distance to the firing equipment it supplies is short or the oil distribution system is very well insulated. Line heaters should be sited close to the burners to raise the oil to its atomising temperature, which should

3/2 A combined steam-electric oil heater with removable elements. The photograph illustrates several desirable features. Thermostatic controls are readily accessible for adjustment, the heater has a clear-dial thermometer, and the whole unit is thermally insulated and clad. For long, trouble-free service, oil heaters should be robustly constructed and electric heating element ratings should not be so high as to cause fuel cracking and carbonisation. 16 kW/ m^2 of heating surface is generally regarded as a safe limit.
A. K. Waugh, Glasgow G3 8PS.

be no higher than necessary to give the oil viscosity specified by the burner manufacturer. Such final preheating of the oil is sometimes carried out within the burner itself.

All handling systems (ring mains and branches) should be completely covered by at least 50 mm (2 in) of suitable insulation, weatherproofed if necessary. If the system is not normally drained on shut-down, or if oil temperatures are liable to drop appreciably during circulation, lines should be fitted with thermostatically controlled tracing. Thermal insulation should not be allowed to come between the oil line and the tracing and any space should be preferably filled by a heat-conducting cement. Particular care should be taken to trace and insulate filter and valve bodies and the condition of insulation and thermostat settings should be checked frequently. Excessive circulation of oil, higher pressures than necessary, over-filtration and dirty filters should all be avoided as they increase pumping costs.

Condensate from oil tank and oil distribution system heating should be reused only if it is regularly inspected and found to be *absolutely* free from oil contamination. No risk should ever be taken. If the condensate cannot be reused it may be possible to pass it through a simple heat exchanger to recover heat before it is discarded.

- **Consult your oil supplier on current oil flow characteristics.**
- **Do not operate delivery line trace heating for longer than necessary.**
- **Always put warm oil deliveries into the tank in current use. Drain off water just before a delivery is due.**
- **Maintain oil in storage at the lowest temperature that will ensure completely free movement to the tank outflow. If possible, raise oil to its handling temperature as it leaves the tank.**
- **Allow reserve oil stocks to cool down, but also allow ample time to bring the fuel slowly up to temperature again when required. Consult your oil supplier first.**
- **To evaluate any proposed insulation, first assess annual average heat losses from oil storage tanks and determine the true cost of steam or electricity to provide this heat.**
- **Reuse condensate from the oil storage and handling system heating only if it is *absolutely* uncontaminated. If it has to be discarded, try to recover the heat.**
- **Reduce oil pumping costs by reducing circulation rates and restrictions in systems.**
- **Ensure that hot oil handling systems and fittings are insulated and where necessary weatherproofed. Tracing should be**

properly fitted and thermostatically controlled, but do not be too lavish with tracing where a small temperature drop can be tolerated.

- **Circulate oil at its correct handling temperature. Do not raise to atomising temperature until just before the burners.**
- **Select the correct atomising temperature to give the viscosity at the burner recommended by the burner manufacturer.**

3.3 Coal

As for other fuels there is no virtue in unnecessarily sophisticated systems and coal and ash handling arrangements should be as simple as possible. Manual labour is expensive, however, and every effort should be made to handle coal

3/3 A well-planned 500 ton reserve stocking area for coal at a large boilerhouse. The concrete-walled bunkers also have concrete bases and there is adequate room for mechanical shovels to manoeuvre. The service bunker is underground, with conveyor elevators to the boilers.
National Coal Board, London SW1X 7AE.

and ash mechanically and to design bunkers and elevator systems to avoid troublesome blockages.

Coal storage requires careful design according to the delivery method, materials of construction and means by which the coal will be withdrawn from the bunker, as well as the type of coal to be handled.

Losses from coal stocks are of some importance. Within bunkers and silos, used for day-to-day requirements, these losses will be small, but coal reserves held in open stockpiles are more likely to deteriorate. Losses into the ground can be reduced by starting with a near-level concrete base (with a slight slope for drainage) or a clean well-compacted foundation of small coal or hardcore. Retaining walls of precast concrete will further reduce the risk of spreading and loss into the ground and on the wind. Walls will also enable more coal to be stockpiled on the same ground area. The open side of any bay should face the prevailing wind.

The exposed surfaces of freshly mined coal tend to oxidise, causing heating and a reduction in calorific value by 3% or so. Most of this surface oxidation occurs in the first four months after mining and it is therefore good practice to retain weathered coal in stock, that is, to adopt a 'first in, last out' system, though all stocks should be turned over periodically.

For small coals (½ in–0), oxidation is usually controlled by compacting stockpiles to exclude air as far as possible. This is less easy for graded coals (1½ in–½ in) and oxidation is usually accepted, steps being taken to limit the resulting temperature rise by controlled ventilation of the stockpile. Temperatures should not be allowed to reach 60–70°C, at which point spontaneous combustion may occur. In larger stockpiles of 200 tonnes or more, temperature measurement may be desirable in the early stages. The coal supplier should be asked to advise on heights and shapes of stockpiles that will minimise oxidation and provide safe conditions.

When handling coal in and out of stock, care should be taken to cause as little size degradation as possible. Anthracites disintegrate very little. Dull hard coals are generally more friable than soft bright ones.

- **Simplify handling as far as possible, but avoid manual handling of coal and ash.**
- **Stock in an orderly way on concrete or hard ground. Use purpose-built silos, bunkers or bays if possible.**
- **Ensure that large open stockpiles are carefully consolidated and that ventilation is controlled to prevent rapid oxidation. Seek expert advice on stacking methods, especially for fine bituminous coal.**
- **Maintain older coal stocks as reserves. Use recent deliveries first. Adopt a 'last in, first out' policy for coal.**

3.4 Natural gas and LPG

Because it cannot readily be stored by many industrial and commercial users, the incidental costs associated with mains gas are minimal. However, where there are interruptible gas supply arrangements, the full cost of standby fuels and their storage should be determined and added to the cost of the gas. The level of stocks of standby fuel must be carefully controlled and the cost of stocking weighed against the probability and effect of an interruption in gas supply.

Natural gas (and manufactured gases) are usually stored only by very large users who may also utilise their own supplies of blast furnace or coke-oven gas. Such users are familiar with the general principles of loss prevention and tracing 'unaccounted-for' gas, but for *all* users regular and systematic checks for gas leaks and the balancing of metered quantities are most important. All reported leaks or escapes of gas, no matter how small, should be investigated at once.

A simple way to check for gas leaks within factories (i.e. beyond the distribution mains, which are normally the responsibility of the supplier) is to turn off all gas-using plant and observe the meter for a period of say 15 minutes. This will indicate if there are any large leaks. Smaller leaks may sometimes be

3/4 Leak-seeking for flammable gases using a probe and portable indicator.
Draeger Safety Ltd, Chesham, Buckinghamshire.

located by smell and can be measured by noting the pressure drop over a fixed time of an isolated length of pipe of known volume. Such a method is described in British Gas Report IM5, 'Soundness testing procedures for non-domestic gas installations'. For larger premises the regular use of a portable combustible-gas detector may be justified.

Liquefied petroleum gases (butane and propane) require special consideration in view of the high capital cost of pressure vessels for bulk storage, and evaporators if offtake rates are high. Usually the prime concern in the case of an LPG leak will be safety rather than the loss of product. Large leaks are usually very obvious because of the 'frosting' which takes place around them.

As butane and propane are of high calorific value and denser than air it is especially important not only to prevent leaks but also to ensure that if one does accidentally occur, no accumulation of vapour can develop in nearby gulleys and drains.

Liquefied petroleum gases are stored in pressure vessels, as liquids. On bulk vessels, relief valves are provided to handle any excess pressures arising from accidental overheating. All vessels have predetermined safe vapour spaces, usually measured with fixed liquid-level gauges, and pressure relief valves are designed for vapour rather than liquid discharge. They are not intended to deal with overfilling, which should *never* be permitted. Accurate stock control is vital.

Every user should be familiar with the recognised codes of practice like those produced in the UK by the Liquefied Petroleum Gas Industry Technical Association and in the USA by the National Fire Protection Association (Code No. 58) and should have a properly laid-down and clearly understood procedure for dealing with LPG leaks and with discharges from pressure relief valves.

- **Include the full cost of standby fuels in the total cost estimates for interruptible gas.**
- **Consider stock levels of standby fuels in relation to the probability and effect of an interruption.**
- **Evaluate carefully the storage and the other equipment costs for LPG as well as its premium qualities.**
- **Check regularly and systematically for gas leaks and deal immediately with all reported leaks and escapes.**
- **Pay particular attention to LPG leaks and escapes.**
- *Never* **overfill LPG storage vessels.**
- **Have procedures ready for emergencies** *before* **they arise.**

4 BOILERS

4.1 Thermal efficiency

The term 'boiler' covers a wide range of plant from simple equipment to provide hot water to very large water-tube boilers of advanced design for power generation. Boilers may also be fired with a number of different fuels using various types of burner. Particular boilers, firing equipment and power systems are beyond the scope of this handbook. Certain broad principles of design and operation apply to them all, however, and these are considered in this section. Many of the points made are related to economic-type steam boilers as most readers will be familiar with this type of plant.

The main contribution to the true cost of steam or hot water at its point of use is the net cost of fuel. Other costs, the fixed and variable overheads listed in Section 2.4, are still significant, however, though often, quite wrongly, ignored. All these costs should be minimised as far as possible and this is greatly aided by attention to two important, interrelated, factors.

The thermal efficiency of the boiler.

Variation in boiler output, or load factor.

Thermal efficiency depends first of all on the design of the boiler and the matching of boiler and firing equipment. If the design is good, a high proportion of the heat produced will be transferred to the water over a range of loads, there will be no operational weaknesses as far as corrosion, cleaning, maintenance and adjustment are concerned and the boiler will be well insulated. Care is needed, however, when considering boiler efficiencies. Test results are usually obtained, quite properly, under steady full-load conditions and take no account of blowdown losses — conditions somewhat different from those which occur in normal use.

Much of the responsibility for high efficiency *potential* lies with the designer but the user still has a considerable influence on the efficiency obtained in *practice*, especially as far as load factor and combustion performance are concerned.

4.2 Load factor and operation

The first objective of energy conservation as far as steam and hot water are concerned will normally be to produce as little of these commodities as

4/1 A well-managed heating plant. The boilers and system are well maintained and both have sufficient instruments to facilitate regular monitoring of performance. *UBH (Mechanical Services) Ltd, London SW4 6RG.*

---

possible; that is, priority should be given to cutting demand. This will certainly be the case, for example, where an existing steam boiler is overloaded. Economies in steam utilisation, reduction of peaks in demand, improved condensate recovery and improved thermal efficiency should all be investigated *before* additional plant is contemplated. In this way the 'energy overheads' will be more satisfactorily spread.

Boilers generally operate at their highest thermal efficiencies when they are neither under- nor overloaded; that is, at about their maximum continuous rating (MCR). Overall efficiency falls off sharply below 50% load as heat losses from the boiler shell, which represent only about 2% of heat input in a modern package boiler at full load, increase as a percentage of heat input and as air control becomes more difficult.

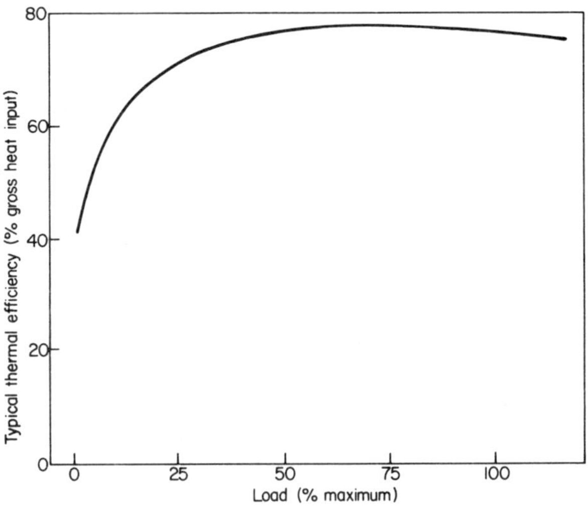

4/2 Efficiency of a shell-type oil-fired boiler at various loads. (After D. C. Gunn.) The overall efficiency of most boilers falls off sharply below 50% of their maximum rating as constant heat losses from outer surfaces become a more significant proportion of heat input and as control of air/fuel ratios and air mixing becomes more difficult.

When selecting new boilers, or when operating existing ones, their numbers and size should be carefully reviewed so that those in actual use can always operate as closely as possible to their designed ratings. Decisions are easier if existing or future load patterns are known on an annual, monthly and daily basis. On existing plant, a load survey will reveal opportunities to reduce peaks in demand. For new plant every effort should be made to obtain realistic load profile forecasts.

Space heating boilers pose particular problems. One boiler should be selected for summer load requirements. It should be sized to meet the load comfortably, but with not too much margin. The development in recent years of modular hot-water boilers and matching sequencing control systems has

4/3 A gas-fired modular boiler, designed to meet the problems of load variation associated with many space heating and hot water applications. Each 50 kW output module is separately controlled and fired, the number in use being matched to load demand to maintain high thermal efficiency.
Stelrad Group Ltd, Hull, North Humberside.

simplified the problems somewhat but many boilers still operate in summer at very low load and hence very low efficiency.

Where an extended heat distribution system is in use it may be worth while to install local services to provide hot water in summer. For example, despite the high cost of electricity, electric storage water heaters may be cheaper to operate than even the smallest fuel-fired boiler if heat losses in distribution are disproportionately high for the load carried.

The necessity for standby capacity should also be examined and the effect in cash terms (and probability of the non-availability of a boiler) should be evaluated. Not to have spare boiler capacity but to ensure that boilers are well maintained so that breakdowns do *not* occur is likely to be more efficient than operating with an ample capacity of indifferently maintained plant.

Boilers in use should be operated as continuously as possible, the maximum firing rate being adjusted to match the maximum demand for heat. This is preferable to frequent on–off operation of a boiler fired at too high a rate. On–off operation aggravates heat losses to cold purging air and, for coal firing, leads to general wastage of fuel during idling periods. Purging must always be sufficient effectively to clear any trace of flammable gases from the boiler and flues but it should not be excessive. National standards and codes (e.g. in the UK, BS 799 Part 4) give guidance on minimum requirements. Users should consider, in order to control heat losses, whether or not the minimum requirements should be greatly exceeded.

Overfiring and blowing safety valves — both wasteful — can be reduced if increases and decreases in demand can be anticipated. When heavy and sudden demands for steam or hot water are unavoidable, large-capacity 'thermal storage' boilers or some other thermal accumulator should be considered as an aid to steadier firing conditions. A sensitive outside air thermostat can be used to anticipate increases or decreases in space heating demand.

Heat losses from boilers taken off load can be markedly reduced if they can be isolated from their flue and chimney system by means of well-fitting gas-tight dampers.

The temptation to make economies by reducing the operating pressure of steam boilers should be resisted. The dryness of steam at the point of use may well suffer and distribution mains may be inadequate. Economies in this direction should be sought by reducing pressure near the point of use to match the optimum steam pressure required for the particular equipment being supplied.

- **Before increasing boiler capacity, seek every opportunity to reduce demand, smooth the load and increase thermal efficiency.**
- **Operate boilers as closely as possible to their designed ratings. Select plant that will enable you to do this throughout the year.**
- **Evaluate the need for standby plant or spare capacity. Better maintenance may be cheaper.**
- **Avoid unnecessary on–off operation by reducing maximum firing rates so that boilers just meet the maximum demands made on them.**

- **Check that purging is not excessive, but do not risk plant safety by purging insufficiently.**
- **Try to anticipate sudden changes in demand. When heavy demands cannot be avoided, use some form of thermal accumulator to smooth firing rates.**
- **Isolate off-load boilers with well-fitting dampers to cut heat losses.**
- **Operate all boilers at their designed pressures.**
- **Consider localised units for hot water services, especially in summer.**

4.3 Combustion performance

Losses during fuel combustion arise directly from:

Incompletely burned fuel, as solids or gases.
Heat in the dry flue gases, including excess air.
Heat in the water vapour of the flue gases.

Incompletely burned fuel

Losses occur when carbon (as smoke, soot or coke), carbon monoxide or volatile components of fuels are emitted and also when combustible materials remain in coal ash. Cenospheres — fractured hollow carbon particles — sometimes arise from the combustion of oil fuels. If modern, well-adjusted firing equipment is used these losses should not be high, but the requirement for a plentiful supply of air to ensure complete combustion conflicts with the need to minimise excess air in order to control SO_3 and NO_x formation and to reduce stack losses. This may be demonstrated in the case of an oil burner, where a marked increase in stack solids (carbon plus ash) emission usually occurs when excess air is reduced to a very low level. Good combustion needs an intimate, uniform and controlled mixing of fuel and combustion air, in the correct proportions in a self-sustaining hot environment. Flame chilling caused by an excess of cold air or impingement of the flame on a relatively cold surface should be avoided. The initial selection of firing equipment and its subsequent maintenance are of paramount importance. Firing equipment that is out of date and cannot meet current requirements should be replaced.

For gas firing, losses from incompletely burned fuel should be negligible. Because carbon monoxide is a potential source of heat loss, as well as being poisonous, levels should be as low as possible — a value of 2000 ppm (parts per million) is unacceptably high. Carbon monoxide is normally measured at the combustion chamber outlet using colorimetric detector tubes.

For oil firing, stack solids emissions of carbon plus ash should not normally exceed 0.4% wt of fuel and with well-designed burners, especially using distillate fuels, values much lower than this should be obtained. Bacharach or Shell smoke scale numbers of 2 or less should be aimed for. A scale number of 4 should be the maximum acceptable for residual fuels.

Losses from coal firing can be very much higher, chiefly from carbon-in-ash, ranging from 2% in a well-run power-station boiler using pulverised fuel to 10% or more on a stoker-fired boiler using a difficult coal. Because of the range of coal and firing equipment available, each application should be individually considered and expert advice on realistic target levels sought from the coal supplier.

Heat in dry flue gases

Heat is carried out of the boiler by carbon dioxide and other oxides formed on combustion and also by nitrogen, most of which plays no significant part in the combustion process. In addition, any air entering the boiler above the minimum (stoichiometric) amount required to provide the oxygen for complete combustion also removes heat. Such heat losses can be serious and they should be controlled by minimising the flue gas temperature as well as the

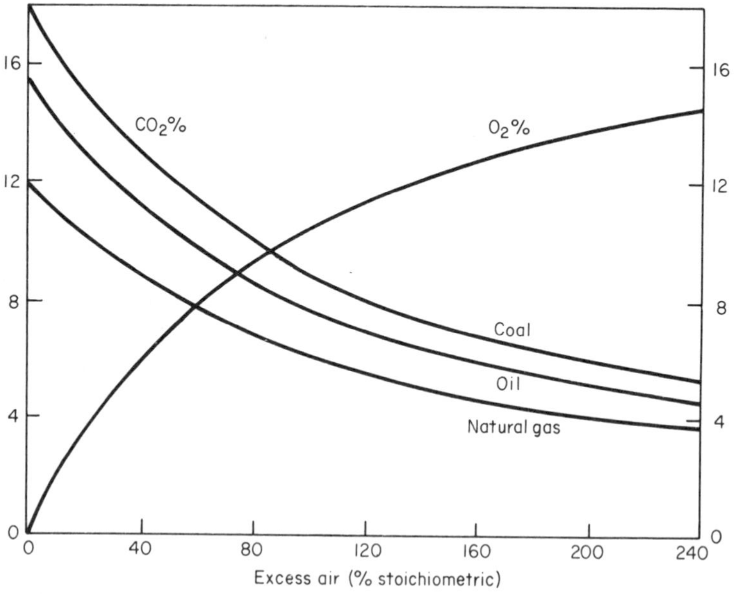

4/4 Relationship between excess air and oxygen content of dry flue gases and between excess air and carbon dioxide content for different types of fuel.

amount of excess air used. Various charts have been devised, for all fuels, relating excess air, flue gas temperature and stack loss.

Excess air. Air infiltration ('tramp' air) should be prevented by sealing cracks in brickwork, round burners, etc. so that *all* air that enters the boiler passes into the combustion zone. Firing equipment should be adjusted throughout its intended operating range to give the minimum air/fuel ratio that will not cause emission problems.

Modulating burners usually require higher excess air levels at lower firing rates to partially compensate for lower air and fuel velocities (which result in less effective air–fuel mixing) and the difficulty of controlling air flow. On oil burners, especially, the maintenance of stable flames poses particular problems at high turndown.

With burner quarls and air registers of fixed geometry, a turndown ratio of, say, 10 : 1 (on any fuel) requires a hundredfold reduction in air pressure in order to maintain a constant air/fuel ratio, a change that is difficult to control, so that burners with very high turndowns invariably run with excess air at low firing rates.

In addition to deterioration due to mechanical wear, fouling (which may decrease air flow rates for the same indicated pressure drop) and changes in fuel composition, any initial setting of air and fuel flow rates is complicated, in multiburner applications, because of inequalities in air distribution.

The performance of the firing equipment depends to some extent on the quality of the air and fuel with which it is supplied.

The general air supply to the boilerhouse must be adequate and air supplied to the burner should be at a constant pressure.

Oil should be of constant viscosity and excessive temperature swings should therefore be avoided. Oil, too, should be supplied to the burner at a constant pressure.

Gas supply pressure should be constant.

Coal should be as consistent a mixture as possible as far as size, ash, moisture, volatile content and reactivity are concerned

Excess air may conveniently be measured by the extent to which it dilutes the flue gases as indicated by their carbon dioxide or oxygen contents.

Flue gas analysis is the basis of simple manual efficiency test equipment (*see* Appendix B). In more sophisticated combustion control systems, changes in air and fuel inputs are also often based on continuous flue gas analysis by various methods. A 3% increase in carbon dioxide level will give approximately a 4% increase in efficiency at the same flue gas temperature. The following are modest target carbon dioxide values for Economic-type boilers operated at full load. Higher values are desirable.

Oil	13%	Natural gas	10%
Coal	12%	Manufactured gas	8%

Flue gas temperature. A flue gas temperature rise of 17°C (30°F) results in a reduction in efficiency of about 1%, at the same excess air level. Temperatures should not therefore be any higher than necessary to prevent condensation at the chimney top, under all load conditions. In practice, for modern boiler plant, the user will usually have to accept the temperature that the boiler provides when it is clean, and ensure that the flue and chimney system used with it is entirely suitable.

Flue gas temperatures can be minimised by *regular* cleaning of tubes and other heat exchange surfaces, or by cleaning as soon as a predetermined temperature is reached, say 20°C (35°F) above the temperature for a newly cleaned boiler. Cleaning must be thorough as gas-side deposits tend otherwise to build up again more quickly. Boiler cleaning should only be undertaken by properly equipped personnel and deposits should not be inhaled or allowed to come into contact with the skin.

Condensation of water vapour in the flue gases from combustion of a sulphur-free fuel occurs at about 45–55°C (110–130°F), the *water* dewpoint. In addition, for a fuel containing sulphur, there will be an *acid* dewpoint usually in the range 115–150°C (240–300°F). There is no significant depression in acid dewpoint temperature until the sulphur content of the fuel is reduced below 0.5% wt. Flue gases from fuels containing sulphur must therefore always be maintained throughout the system above a minimum temperature of 150°C (300°F) to prevent any possibility of acidic condensation.

Because of differences in heat transfer characteristics between gas and oil flames, the latter being more radiant, a conversion from oil to gas may produce significant changes in gas temperatures within the boiler, resulting in the overheating of tube plates and tube ends. The possibility of this sort of damage occurring and of higher flue gas temperatures, should always be considered when any change in fuel is contemplated.

Where flue gas temperatures are sufficiently high, heat may be recovered by preheating combustion air or feedwater. The installation of air heaters and economisers should be undertaken only after detailed study and on expert advice, if corrosion and other problems are to be avoided. Existing heat recovery equipment should be maintained in good order and operated correctly. In assessing boiler efficiency, allowances must be made for air or feedwater preheat.

Heat in water vapour

Free moisture with fuels should be reduced as far as economically possible before firing, provided that, for coal, handling properties are not impaired. Nothing can be done, however, about inherent moisture in the fuel or the water

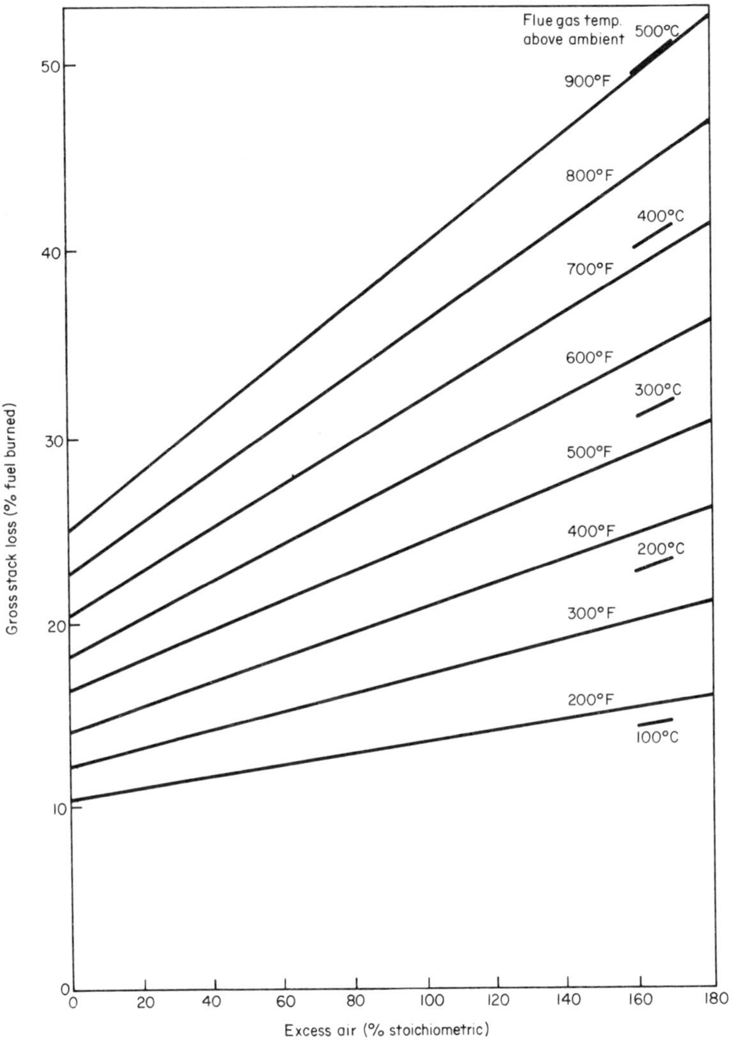

4/5 Heat loss in flue gases (stack loss) when using a residual oil fuel, as a percentage of gross input. An approximate indication of boiler efficiency may be obtained by measuring the heat loss in the flue gases and deducting this (plus an estimate of the radiation loss from the boiler) from 100%. The stack loss depends on the flue gas temperature above combustion air temperature, the amount of excess air (% of stoichiometric) and on the specific heat of the flue gases, which varies with the composition of the fuel used. Various charts and formulae are available, with stack losses related to net or gross heat input. They assume that all the fuel is burned; if it is not, allowance must be made for this.

vapour produced by the combustion of hydrogen and hydrocarbons.

The loss of the latent heat of vaporisation of water (i.e. the additional heat energy required to change water from liquid at 100°C to steam at the same temperature) amounts to about 2.3 MJ/kg (1000 Btu/lb) of water. The highest losses due to water vapour occur with gaseous fuels, the lowest with dry anthracite. They are reflected in the hydrogen contents and in the difference between gross and net calorific values in each case.

- **Avoid using fuels with large amounts of free moisture. Drain water from coal bunkers and oil storage tanks.**
- **Seal all cracks and holes around burners and boiler flues to prevent air infiltration. Seal smoke-tube doors.**
- **Keep all air and fuel supplies to the firing equipment as constant as possible. Avoid fluctuations in pressure and composition.**
- **Reduce losses from unburned fuel by improving firing conditions, but avoid using more air to do so. Look for flame chilling, poor atomisation, poor mixing and other causes of incomplete combusion.**
- **Reduce excess air and improve efficiency by using the lowest air/fuel ratios that will not cause combustion problems with the equipment you are using. If excess air cannot be reduced to an acceptable level, overhaul or change the firing equipment.**
- **Ensure, under all firing conditions, that condensation will not occur in flues and chimneys. Maintain a minimum flue gas temperature of 150°C (300°F) at the chimney top for all fuels that contain sulphur.**
- **Clean tubes and other fire-side heat exchange surfaces regularly or as soon as a predetermined temperature is reached. Cleaning should be thorough.**
- **Maintain air heaters and economisers in good order and operate them correctly. Keep a close watch for signs of corrosion.**

4.4 Boiler water (*see also* Section 10.3)

Correct feedwater treatment is necessary in steam boilers to prevent water-side deposits which reduce heat transfer rates and lower efficiency, especially in modern compact units where heat fluxes are very high. The level of solids which concentrate in the boiler water is controlled by discharging some of it to waste, periodically or continuously. Heat losses due to this blowdown can be reduced by several methods.

Not blowing down any greater volume than necessary to maintain the total dissolved solids (TDS) below the level recommended by the boiler manufacturer.

Reducing the amount of blowdown required by recovering a high proportion of clean condensate, thus reducing the amount of fresh make-up water required.

Recovering heat from the blowdown partly as flash steam (*see* Section 5.5) and partly via a heat exchanger. The latter may be economic only on a large installation. Preheating feedwater is a common means of recovery.

Careful control of feedwater treatment, which should be no more than necessary.

Losses from blowdown can easily amount to 3% of the boiler heat input so that any reduction or heat recovery is likely to be worth investigating. The utmost care is necessary in making any modifications to blowdown systems and the work should only be entrusted to an experienced contractor. Pipework must be properly secured. Fatal accidents have occurred owing to pipes being unscrewed by the considerable turning moment which can be produced by an issuing water jet.

- **Ensure that correct feedwater treatment is carried out. Meter the feedwater and do not overdose.**
- **Do not blow down any greater volume of boiler water than necessary and recover the heat if possible.**
- **Reduce the amount of make-up water (and hence blowdown) by recovering as much clean condensate as possible.**

4.5 Control, maintenance and training

Good housekeeping and adequate boilerhouse records are essential if high thermal efficiency is to be obtained. BS 1374:1972 gives recommendations and specimen log sheets. Fuel use over fixed periods should be compared with steam output, ambient temperature or other appropriate parameter. Steam should be metered and not just inferred from manufacturers' data for steam-heated equipment. Overall load factor should also be determined, the total steam generation or hot water output over an appropriate fixed period being compared with the total capability of the steam-raising plant actually in use. Manufacturers' advice should be sought on the accuracy of meters.

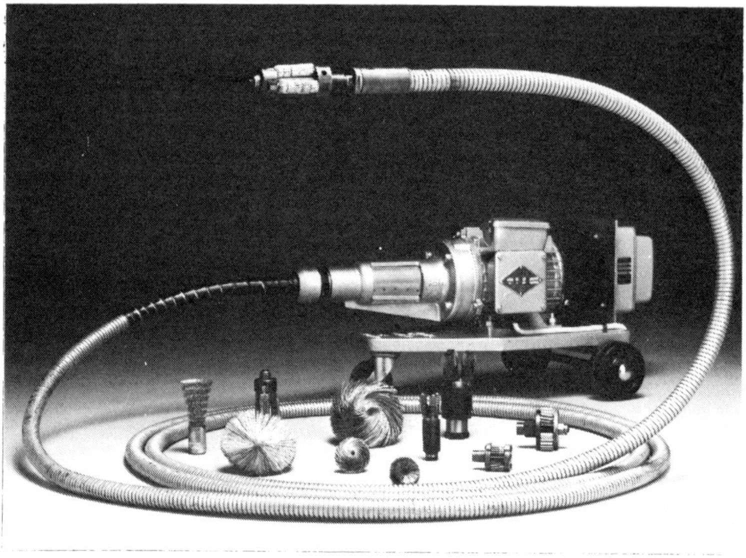

4/6 Soot and scale are good thermal insulators and high boiler efficiency can only be maintained by regular and thorough cleaning of fireside surfaces. A mobile flexible power drive and selection of toolheads and brushes to deal with all types of deposits. *Flexible Drives (Gilmans) Ltd, Warwick.*

Measurements should be made periodically on boilers, to determine efficiencies and to identify where improvements are possible, using portable test equipment. Any full-scale efficiency tests should be to a recognised standard procedure (e.g. BS 845:1972), but quick and useful results may be obtained by examining the principal losses from the boiler rather than by carrying out a full heat balance.

Stack losses as a percentage of heat input (net or gross) may be determined from measurement of the carbon dioxide or oxygen content of the flue gases, together with flue gas temperature. Samples must be representative and should be taken under steady conditions. Temperature measurements taken with thermocouples or mercury-in-glass thermometers tend to be low. An approximate correction may be made by adding 10% to the reading. More reliable values may be obtained by using a suction pyrometer.

A permanent combustion test point should be fitted to each boiler flue outlet and sealed when not in use. It should be well clear of any bend, junction or obstruction in the flue and there should be no air leakage round the probe. Measurements are more likely to be reliable if the conditions for the tester are reasonable. There should be proper illumination, space in which to work and somewhere to stand instruments.

High efficiency cannot be maintained without regular servicing and adjustment of firing equipment and controls. All measuring devices should also be calibrated periodically. A servicing schedule should be drawn up with special emphasis being given to those aspects which affect thermal efficiency and safety. When maintenance is undertaken by outside organisations (often, the equipment suppliers or installers) a clearly defined and agreed programme of work is still needed.

In addition to any servicing schedule, a copy of the standard operating instructions should be kept near the burners. Essential instructions, flow diagrams and wiring diagrams (suitable for the operator, as distinct from design drawings) should be prominently displayed and a systematic method of tracing burner faults should also be readily available, if possible.

Maintenance is greatly facilitated by the provision of adequate working space with good illumination near the burners and safe access to all items requiring attention. Valves, controls, motors, starters and pipes should be clearly and permanently identified.

A build-up of carbon or damage to any part of a burner nozzle is likely to result in poor combustion. The correct positioning and condition of atomisers, gas spuds, air guides and ignition equipment should receive particular attention. Pressure-jet oil atomisers are especially vulnerable items. They are subject to wear in use and should always be protected by fine filters. Smaller sizes are difficult to clean and should be replaced periodically by atomisers of the same rating, spray angle and spray pattern. Manufacturers provide useful guidance on selection and maintenance. Any cleaning that is undertaken should be carried out with the utmost care. Metallic scrapers and abrasive cloths should not be used. The atomiser should first be soaked in gas oil and then cleaned with a soft cloth or soft *wooden* scraper.

All persons involved with the operation and maintenance of boilers should be adequately trained, the importance of burner operation in relation to overall plant operation and thermal efficiency being properly emphasised.

A most useful training opportunity occurs during plant commissioning and whenever possible, operators should be present. This period is also a useful one for highlighting additional training needs for those with some basic knowledge but, perhaps, limited experience. Nevertheless, the need for subsequent training, especially of new operators, should not be overlooked.

Ideally, operator training should be divided between off-the-job instruction in basic combustion, purpose of the equipment and procedures, and on-the-job training conducted by an experienced instructor. Various organisations have standard training programmes for boiler operators and the City & Guilds of London Institute provides examinations for its Boiler Operators' Certificate.

Training should be repeated periodically to update operators' knowledge, particularly when new equipment is introduced. Refresher courses have the added advantage of challenging existing practices, especially if conducted by someone from outside the organisation.

- **Maintain simple boilerhouse records in order to identify trends and changes in boiler conditions.**
- **Regularly compare fuel used with steam or hot water output. Try to improve the ratio.**
- **Check load factor periodically by comparing, over a fixed interval, total steam raised (or hot water produced) with the total capability of the plant used.**
- **For full-scale efficiency tests, follow a recognised standard procedure. If necessary, get expert advice, at least for the first few tests.**
- **Carry out short tests of combustion performance at regular intervals. Measure the carbon dioxide (or oxygen) content of the flue gases and the flue gas temperature. From these values assess the stack loss, using charts appropriate to the fuel used.**
- **Make testing easier and more reliable by fitting proper test points.**
- **Regularly service all firing equipment and calibrate instruments and other measuring devices. Pay particular attention to atomisers.**
- **Ensure that boiler operators are adequately trained and properly instructed in their duties.**

5 STEAM AND HOT WATER UTILISATION

5.1 Heat distribution and insulation

There is little point in producing heat at high thermal efficiency — in steam, hot water or heat transfer fluids — if the effort is subsequently squandered in poor distribution and utilisation. Overall efficiency is the *product* of individual efficiencies throughout the system.

The potential for heat loss is great in all types of distribution system. They need regular inspection and maintenance to predetermined schedules which should include all valves, traps, gauges and joints as well as the pipes themselves. Particular attention should be paid to the rectification of leaks, which should be regarded as sources of heat and cash loss rather than mere nuisances.

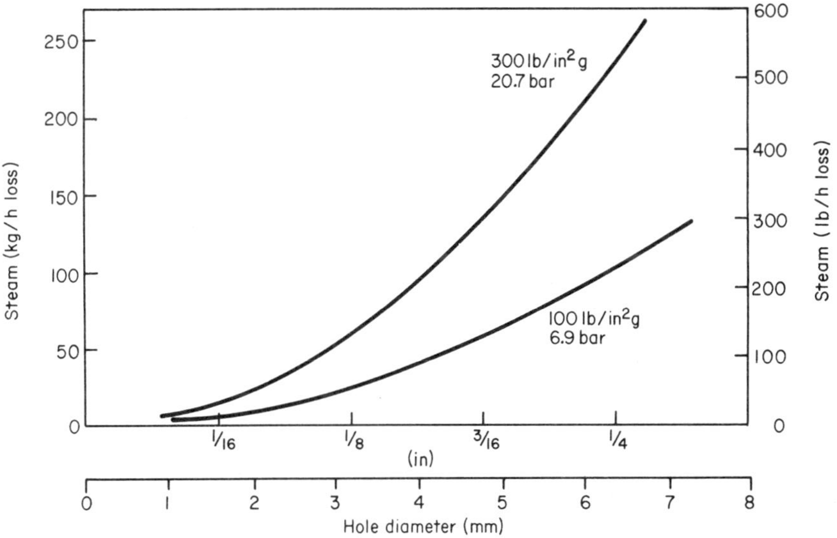

5/1 Steam losses from leaks obviously increase with pressure. They rarely occur through circular holes, but the chart indicates the approximate magnitude of losses. Badly mating flanges are a common source of steam leakage.

All aspects of heat distribution and use need to be reviewed periodically. New equipment may have been added or the conditions of use changed. It should not be assumed that any system once designed and installed will continue to operate at maximum efficiency for ever without some continuing effort. Whenever feasible, out of date or unnecessarily extended distribution systems should be eliminated or at least isolated effectively. Too much reliance should not be placed on old isolating valves. Parallel runs of piping should be replaced by correctly sized single runs.

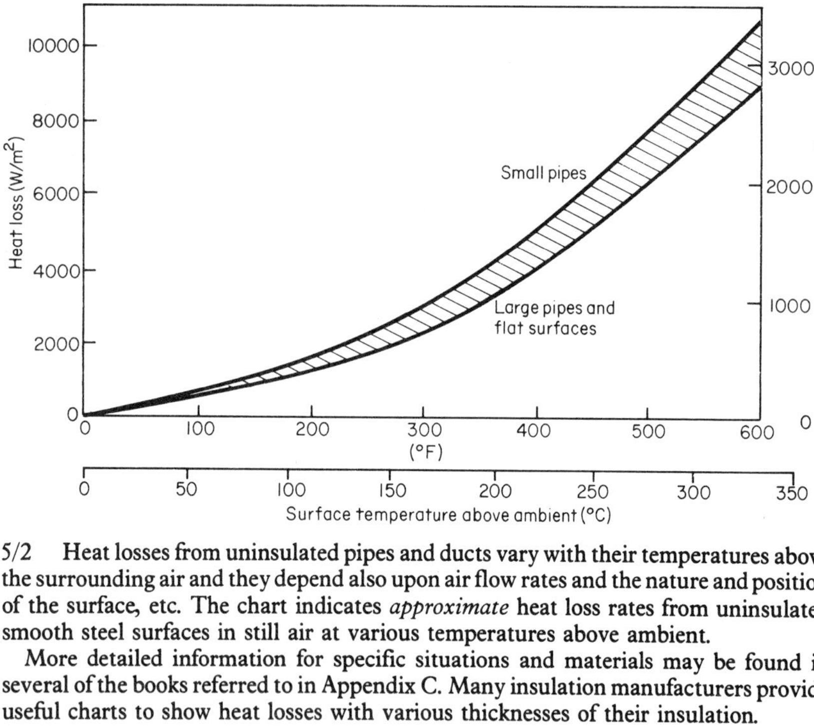

5/2 Heat losses from uninsulated pipes and ducts vary with their temperatures above the surrounding air and they depend also upon air flow rates and the nature and position of the surface, etc. The chart indicates *approximate* heat loss rates from uninsulated smooth steel surfaces in still air at various temperatures above ambient.

More detailed information for specific situations and materials may be found in several of the books referred to in Appendix C. Many insulation manufacturers provide useful charts to show heat losses with various thicknesses of their insulation.

The effective insulation of all hot pipework and surfaces is of paramount importance and tanks, calorifiers, heat exchangers, valves and flanges should all be included. The assessment of economic levels of insulation should be based on a minimum total cost approach (*see* Figure 7/3) but *all* hot surfaces should have *some* insulation. The selection of a new insulating material or a replacement for old lagging should be made only after carefully listing the precise requirements of the application to ensure that the material will be entirely suitable and also sufficiently durable and safe. Once applied, thermal insulation must be kept dry and protected against mechanical damage if it is to continue to function correctly.

Sections of distribution pipework and ducts left deliberately without insulation to provide space heating should be examined critically to ensure that they *really do* contribute heat effectively and in a sufficiently controlled way. Uninsulated, high-level steam pipes in constant use throughout the year are a particularly dubious method of space heating.

As well as having insulated sides and bases, open-topped tanks containing hot materials should, whenever feasible, also be fitted with suitable non-corroding covers to minimise heat and evaporative losses. Ready access can be preserved by the use of a blanket of floating plastic balls. Even a single layer will reduce heat losses by over two-thirds and the method is suitable for a wide range of hot liquids.

- **Develop regular inspection and maintenance schedules for all forms of heat distribution. Include joints and fittings (particularly steam traps).**
- **Correct all faults in heat distribution systems – particularly leaks – as soon as they are noted.**
- **Simplify all heat distribution systems and remove or seal off sections no longer in use.**
- **Insulate all hot pipes and other surfaces and ensure that the insulation does not become crushed or waterlogged.**
- **Ensure that sections of pipe deliberately left uninsulated to provide space heating do so in a controlled and effective way, and in winter only.**
- **Insulate all tanks, vats and other process vessels containing hot liquids. Fit open-topped tanks with covers or use plastic ball blankets.**

5.2 Steam properties

An appreciation of the principal properties of steam and some familiarity with steam tables are desirable for anyone concerned with steam distribution and utilisation. Steam tables are available in various units and degrees of complexity but a simple abridged version is often more than adequate.

Water is raised to its boiling point by increasing its *sensible heat*. The additional heat required to turn water at its boiling point into steam at the same temperature is known as its *latent heat*. In more scientific terms, the heat content of water per unit mass (kJ/kg), related to a defined base, is known as its

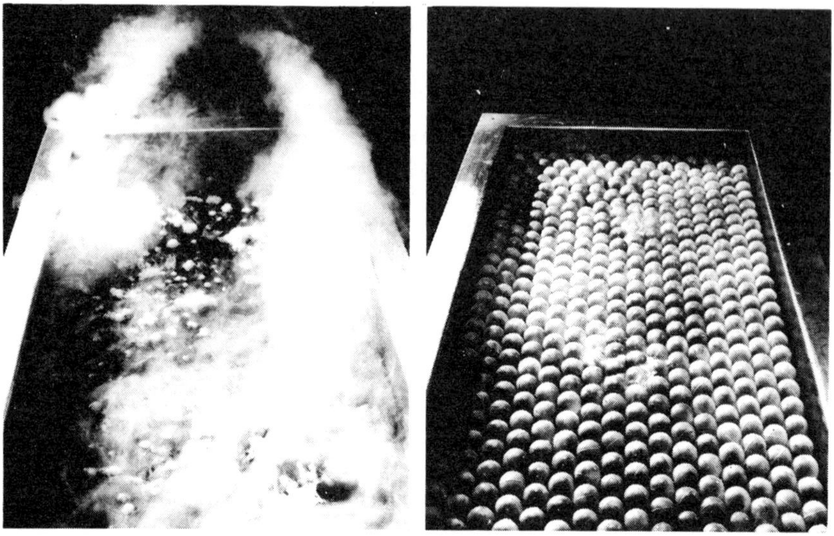

5/3 Before and after. A layer of hollow plastic balls used to reduce fumes and heat losses from a tank of hot acid. The balls form themselves naturally into a single layer with over 90% surface coverage. They are readily pushed aside to give access to the tank and reform the blanket as soon as work is removed.
Capricorn Industrial Services Ltd, London SW1A 1JT.

5/4 Steam properties.

The boiling point of water increases with pressure.

For dry saturated steam $h_g = h_f + h_{fg}$

For wet steam $h = h_f + xh_{fg}$

where x = dryness fraction = a/b

For superheated steam, $h = h_g + C_p\,(t - t_s)$

Where C_p = specific heat at constant pressure.

Steam tables provide extensive data on the specific enthalpy (heat content per unit mass) of steam for various conditions of pressure and temperature, as well as other properties. They are available in SI and Imperial units.

The table opposite provides selected values of specific enthalpy to indicate the relative importance of the specific enthalpy of water ('sensible' heat) and change of phase or evaporation ('latent' heat) at various pressures.

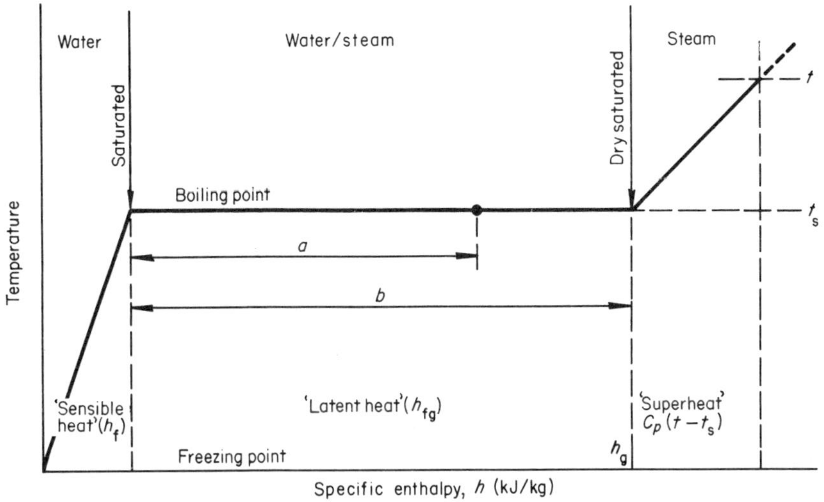

Gauge pressure		*Temp*	*Specific enthalpy*			*Specific volume*
			Water h_f	*Evap'n* h_{fg}	*Steam* h_g	
(bar)	(kN/m^2)	(^{o}C)	(kJ/kg)	(kJ/kg)	(kJ/kg)	(m^3/kg)
0	0	100	419	2257	2676	1.67
2	200	134	562	2163	2725	0.60
4	400	152	641	2108	2749	0.37
8	800	175	743	2031	2774	0.21
16	1600	204	872	1924	2796	0.12
32	3200	239	1034	1770	2804	0.06

specific enthalpy, h. The specific enthalpy of saturated steam h_g is the sum of the specific enthalpy of the saturated liquid ('sensible' heat) h_f and the specific enthalpy of change of phase from liquid to vapour at constant pressure ('latent' heat) h_{fg} i.e.

$$h_g = h_f + h_{fg}$$

Steam is a valuable heat transfer medium because it can convey up to 5½ times its sensible heat content as latent heat which it can then surrender without any reduction in temperature. However, as steam pressure is increased and the boiling point of the water is raised, the specific enthalpy of change of phase, h_{fg} ('latent' heat) decreases. As h_{fg}/h_g decreases, utilisation tends to become less efficient unless positive steps are taken to recover heat in the condensate, h_f. This is considered in Sections 5.4 and 5.5.

Saturated steam is used for most process and space heating applications. Wet steam, that is, steam which contains minute water particles, conveys less heat per unit mass than dry steam, which is therefore usually preferred. Steam with a dryness fraction of say 0.95 will contain 5% by mass of water and its latent heat will be lower by this proportion. Superheated steam, i.e. steam heated above its dry-saturated condition, is used mainly for power generation where high pressure and temperature are essential for reasonable efficiency.

5.3 Steam distribution

The most suitable steam distribution pressure requires careful assessment in each case. There are no invariable rules and the important thing is that the whole plant or system should be considered. A particular pressure should not be selected simply because the boiler happens to provide it or because of the requirements of just one item of plant or equipment. Whenever pressure reduction occurs it is desirable that work should be done or some useful result (e.g. improved dryness) should accompany the change.

Boilers should generally be operated at their designed pressures if priming and wet steam are to be avoided. Subsequent distribution at high pressure requires smaller mains and, by local pressure reduction, the correct pressure for each item of plant and dry steam can more readily be assured. Insulation costs and heat losses will also be lower. On the other hand, steam leaks are more serious at high pressure and there is a danger that individual items of plant will be operated at high pressure simply because it is available and not because it is necessary. Even with simple pressure reduction at the boiler this misuse can still occur, the distribution pressure dictated by the highest pressure required at one point being used for every other item of plant.

Pressure-reducing valves should always be set to provide a pressure at each point of use that is no higher than absolutely necessary. As suggested earlier, it is desirable that work should be done in the process of pressure reduction. Power generation using steam turbines is widely practised by larger industrial users where the initial capital cost can be justified and where a suitable balance of electrical load and process heat exists. Steam pressure may also be reduced by using it to drive air compressors. These are specialist matters which require detailed and careful evaluation of the power and heat balances involved.

Where a widely fluctuating steam load exists, the total boiler capacity can be reduced and boilers operated more steadily, and hence more efficiently, by using an accumulator, which is a pressure vessel containing water into which, at times of low demand, surplus steam is fed to raise its temperature. At peak demand times, the hot water provides steam at a lower pressure for suitable process uses.

To avoid erosion in a system, steam velocities should not normally exceed about 40 m/s (130 ft/s). Somewhat lower velocities are desirable in most systems, however, in order to minimise pressure drops. Precise values of pressure drops can be calculated from data to be found in various standard reference books.

- **Review steam distribution periodically and make adjustments to meet new operating requirements.**
- **Determine optimum steam distribution pressures according to the requirements of the units served, but operate the boiler at its designed pressure.**
- **When using saturated steam, ensure that it is as dry as possible.**
- **Where large amounts of high-pressure steam are used for low-pressure applications, consider how useful work may be done in pressure reduction.**
- **When the load on the boiler has heavy peaks, consider the use of a steam accumulator.**

5.4 Steam utilisation

In addition to insulating distribution systems, preventing leaks and instituting a proper system of general maintenance, attention should be paid to the other principal ways of cutting costs by:

Reducing the amount of heating that the steam is required to do.

Recovering as much heat as possible, especially in higher-temperature applications.

Steam requirements should be closely examined and challenged for each application and optimum processing times and temperatures determined. Rounded figures should be treated with suspicion. Provided process times are not unduly extended as a result, lowering process pressures and temperatures usually results in lower heat losses and also in reduced losses from flash steam. Whenever possible, temperatures should be closely controlled using correctly installed thermostatic valves.

Steam requirements may also be reduced by:

Improving heat transfer rates by stirring or agitating the materials being heated and keeping heat exchange surfaces clean.

Using dry steam and draining condensate effectively as soon as it forms.

Loading process plant to its designed capacity but not overloading it. If possible, the number of items of plant in use should be reduced.

Preheating incoming material using heat recovered from some other process.

In drying processes, removing water by mechanical means such as spin drying or allowing more time for natural drainage before processing.

Recording and reducing the amount of waste or reprocessing, particularly where spoiled material is easily recovered and reprocessing therefore easily goes unnoticed.

Avoiding overprocessing. Materials should be removed immediately they reach the desired temperature or condition.

When live steam injection is practised, analysing the true cost of this in terms of heat and treated water and evaluating possible alternative methods (e.g. indirect steam heating with thermostatic control or direct-fired immersion tubes).

At higher steam pressures and temperatures the ratio h_{fg}/h_g (latent heat to total heat) decreases and heat transfer becomes less efficient. For example, for a process vessel at 100°C supplied with dry saturated steam at atmospheric pressure (0 bar g) (see Figure 5/4):

$$h_g = 2676 \qquad h_{fg} = 2257 \qquad h_f = 419 \qquad \text{(kJ/kg)}$$

h_f represents the heat remaining in the condensate at 100°C, h_{fg} the heat available to the process. The best possible thermal efficiency is therefore

$$\frac{2257}{2676} \times 100 = 84\%$$

For a similar vessel operated at 239°C supplied with dry saturated steam at 32 bar g:

$$h_g = 2804 \qquad h_{fg} = 1770 \qquad h_f = 1034 \quad \text{(kJ/kg)}$$

In this case, the best possible thermal efficiency is

$$\frac{1770}{2804} \times 100 = 63\%$$

As steam process temperatures and pressures are increased, heat recovery from the process condensate and, in evaporators, from the evaporated water, becomes *essential*.

- **To reduce steam requirements, determine for each item of steam-heated plant the optimum steam pressure, process temperature and time. Control these factors automatically.**
- **Load process units to capacity — but do not overload. Reduce the number of units in use.**
- **To maintain high heat transfer rates, use dry steam, drain condensate quickly, keep heat exchange surfaces clean and heated materials agitated or stirred.**
- **Preheat materials with recovered heat if possible or dry them partially by non-thermal methods.**
- **Cut down waste, reprocessing and overprocessing.**
- **Consider alternatives to direct steam injection for heated tanks and vats.**
- **As steam temperatures and pressures increase, pay more attention to heat recovery.**

5.5 Heat recovery from steam

Condensate

Uncontaminated steam condensate is a valuable source of heat and pure water and it should be recovered whenever it can usefully be used. A 6°C (11°F) increase in boiler feedwater temperature is approximately equivalent to a 1% reduction in fuel cost. Because it increases the temperature of the boiler feedwater, it also increases the effective boiler capacity at no additional cost, an important consideration when a boiler appears unable to meet the demand for steam. Close control of boiler feedwater is essential. The volume and temperature of returned condensate should be recorded together with the volume of make-up water and the feedwater temperature. Condensate recovery is an opportunity to economise on water treatment and only the make-up water should be treated, not the condensate, subject, of course, to the regular checks of water quality essential in operating modern boiler plant.

Condensate having been recovered from steam traps, care should be taken to avoid contamination or any significant heat loss from it. Return lines and storage arrangements must be clean and insulated. Condensate storage tanks should be covered, of sufficient capacity and controlled in such a way that overflows do not occur with normal load fluctuations.

Efficient condensate return can result in high boiler feedwater temperatures which may cause pumping difficulties. These may be dealt with by making greater use of flash steam and thus returning condensate at lower temperature, or by using a higher-level feedwater or service tank. The head required will depend upon the pump design, but must exceed the net positive suction head (NPSH) specified by the manufacturer for the water temperature.

But there is little merit in returning large volumes of hot condensate if these arise simply because the distributed steam is wet or because the plant in which it is used is so inefficient that it requires much more steam to do the job than is really necessary. Condensate volumes and temperatures should always be reduced by efficient operation before attention is focused on recovery.

Condensate may be lifted by an ordinary trap using steam pressure in the plant being drained, but it is often more satisfactory to collect it in a vented receiver and pump it into a return main using a steam-operated pumping trap. For larger recovery operations condensate may be returned at high temperature, without flash steam losses and with air separation using a motor-driven booster pump system.

Flash steam

Steam is generated when hot water (e.g. condensate) under pressure is released to a lower pressure. Not only is this a nuisance, but heat is also lost unless positive steps are taken to recover it. The generation of this flash steam is increased and so is the amount of condensate to be handled when process steam is not dry.

Flash steam is not caused by faulty steam traps, although traps are often blamed. Its recovery is unlikely to be practical unless several points at which pressure is reduced can be brought together (e.g. condensate return lines at the hot well), or one large item of plant justifies individual treatment. Flash steam may be condensed by a cold water spray, the temperature of the water being raised, or it may be used directly in a lower-pressure/temperature application, e.g. preheating. As for all forms of recovered heat, the operation is pointless if there is no use for it, or if it merely causes an equivalent waste of heat elsewhere.

Steam traps

Heat transfer is generally improved if air is eliminated by fitting steam mains and steam spaces with automatic air vents. Condensed steam should also be quickly and effectively removed from the steam space. Condensate is usually drained through a steam trap, which is a device that retains steam in the system but allows condensate to pass out in a controlled way. Traps are an essential

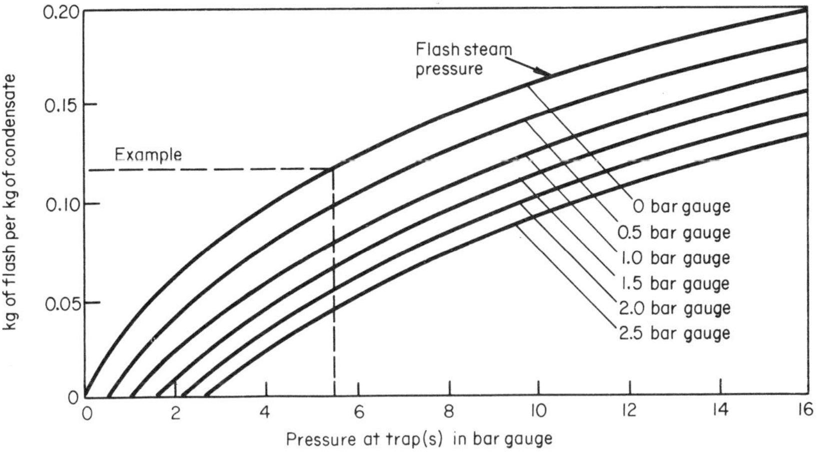

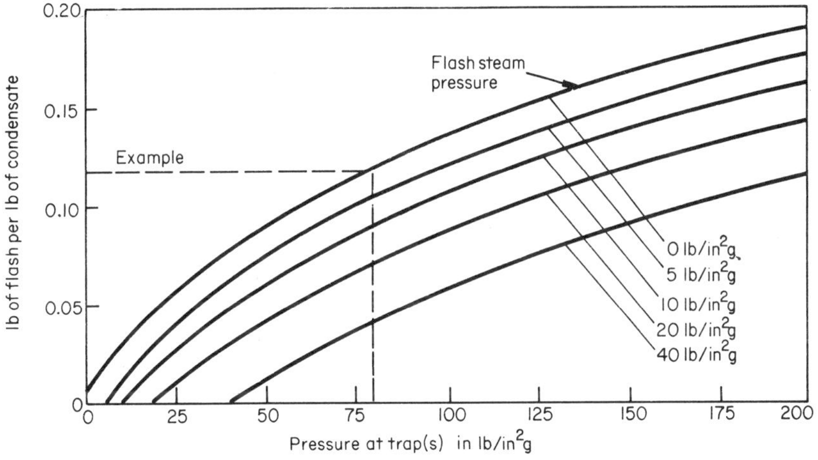

5/5 The amount of flash steam produced for a given weight of condensate depends on the pressure at the trap and also the pressure at which the flash steam is to be used.

If hot water at 5.5 bar gauge (80 lb/in^{2}g) is released to atmospheric pressure, the water temperature drops from the saturated steam temperature of 162°C (324°F) to 100°C (212°F). Excess heat in the form of steam is flashed-off at the rate of 0.12 kg/kg (lb/lb) of condensate. Smaller amounts are available at higher flash steam pressures.
Spirax Sarco Ltd, Cheltenham, Gloucestershire.

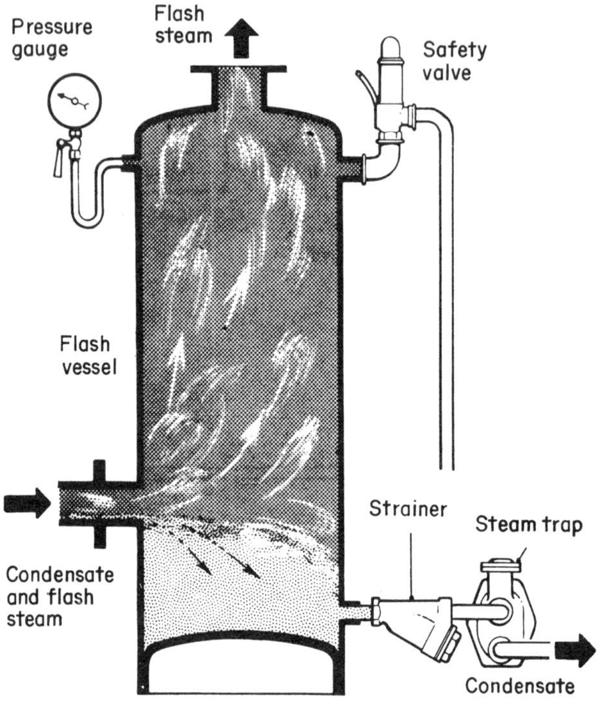

5/6 When a line carries both condensate and flash steam, the flash can be separated and used locally by passing the mixture into a properly designed flash vessel of adequate height. As condensate and flash steam enter, their velocity is sharply reduced, the condensate falling to the bottom where it is recovered through a float-type trap. *Spirax Sarco Ltd, Cheltenham, Gloucestershire.*

---

part of any steam system and they should not be regarded as unfortunate appendages. There are a number of different types available, working on various principles, and reputable manufacturers provide ample guidance on their selection and application.

Sizing should be based on the amount of condensate to be passed and the differential pressure across the trap and *not* on the size of the pipe to which it happens to be connected. Fitting is as important as the correct selection of a trap, and strainers, sight glasses and check valves may also be required to ensure proper functioning. Traps should normally be fitted close to the equipment which they serve and below the natural drain point. They should be indoors whenever possible, protected against frost and readily accessible for maintenance and cleaning. Bypasses should be fitted only if they are essential for

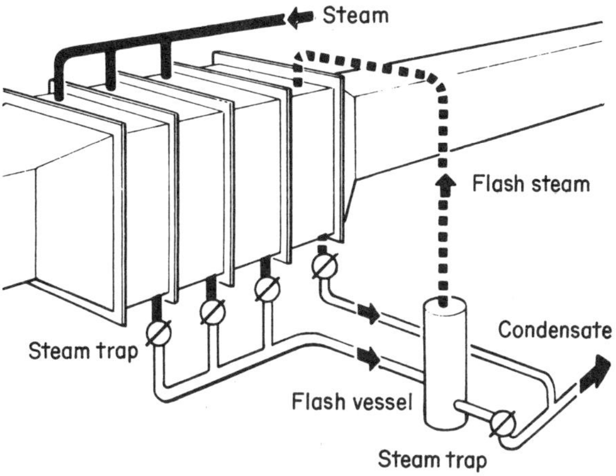

5/7 An example of the good local use of flash steam in an application where its pressure is unimportant. Three air heater batteries use steam at 8.0 bar gauge (120 lb/in^{2}g). The flash steam which they produce can, after separation, be used at atmospheric pressure in an additional preheat battery.
Spiras Sarco Ltd, Cheltenham, Gloucestershire.

maintenance purposes or for warming-up periods. Generally, they should be avoided, as their use provides too ready an alternative to good maintenance.

Group trapping, that is, the use of a single large trap to serve several units, may facilitate flash steam recovery but it should only be employed where load and pressure conditions are constant, and the flow of condensate from any one unit into the common trap will not be interfered with in any way — a situation which rarely occurs in practice.

- **Ensure that recovered heat can be used before recovering it.**
- **Recovered condensate saves valuable heat and chemicals. Keep it hot and clean, but cut the amount produced before concentrating on recovery.**
- **Record the volume and temperature of recovered condensate, the volume and temperature of make-up water and the temperature of the feedwater. When changes occur, investigate the reasons.**
- **Maintain close control of boiler feedwater quality. Make allowance for clean recovered condensate and do not over-treat with chemicals.**

- **Pay particular attention to condensate and feedwater tank insulation. Fit covers. Ensure tanks are high enough to prevent pumping problems and large enough to avoid overflows.**
- **Recover flash steam whenever there is a use for it in a lower-pressure application like preheating. Initially, concentrate on a few good larger schemes.**
- **Encourage everyone concerned to appreciate the purpose of steam traps and ensure that they are correctly selected, fitted, protected, cleaned, maintained and cared for.**

5.6 Hot water distribution and utilisation

A principal use for hot water is for space heating, a topic covered in Chapter 6. A number of suggestions made already regarding steam utilisation apply equally to the distribution and use of hot water for process heating. Particular attention should be paid to the assessment of optimum process times and temperatures.

The use of hot water should be considered for all low-temperature steam applications, particularly where surplus hot water is available or could be obtained from flash steam condensation. When large intermittent demands for hot water for process use occur, adequate hot water storage will smooth boiler operation, but care should be taken to ensure that tanks are well insulated.

Control of the use of hot washing water is essential and sprays and time controls should be used. If hot water can be recycled, after settling and filtration if necessary, this should be done. Where there is no alternative to discharging to the drains, consideration should be given to heat recovery through a simple heat exchanger.

- **Consider hot water heating for all low-temperature steam applications, especially if surplus hot water is available or there is flash steam which can be condensed.**
- **Determine for each item of plant the optimum process temperature and time. Control these automatically, if possible.**
- **Load process units to capacity — but do not overload. Reduce the number of units in use.**
- **Preheat materials with recovered heat, if possible, or partially dry them using non-thermal methods.**
- **Cut down waste, reprocessing and overprocessing.**
- **Keep heat exchange surfaces clean and heated materials agitated or stirred.**

- **Carefully control the use of hot washing water and reuse it whenever possible.**

5.7 Liquid-phase heating

As a heat transfer medium, steam has certain very useful properties, already described, but to attain high temperatures, using saturated steam, high pressures are also necessary. A temperature of 250°C (480°F) requires a saturated steam pressure of nearly 80 bar (600 lbf/in^2 g). This, of course, increases the cost of the system.

Liquid-phase thermal fluids are mainly stable mineral oils and synthetic materials which can be heated to high temperatures but at atmospheric or low pressure in systems similar to hot water systems. They can, of course, only be used for indirect heating. Temperatures up to 500°C (930°F) can be readily obtained with a number of fluids and higher temperatures are possible using molten metals.

Higher thermal efficiencies may be obtained using thermal fluids in applications like indirectly heated ovens or in place of direct-fired tank heaters, but the main cost-saving features of liquid-phase heating lie in its overall simplicity. Advantages claimed are:

Simpler distribution pipework, with fewer valves, traps, etc.

Higher availability, through lower corrosion and more uniform heat transfer.

No feedwater treatment, blowdown losses or condensate recovery.

Quick warm-up and rapid response.

Higher operating temperatures.

Lower labour costs and reduced statutory requirements.

A change of heating method to liquid phase requires careful evaluation because, in making cost comparisons, a number of additional and not readily quantified factors are involved.

6 INDUSTRIAL SPACE HEATING

6.1 Heating requirements

Too little consideration is given to the needs which must be met by industrial space heating, these needs generally differing considerably from those to be met by domestic heating systems. Industrial heating may be used to provide comfortable conditions for people doing a variety of jobs, but also for such purposes as the correct storage of goods or the provision of precise environmental conditions for sensitive equipment. Heating is not the only environmental requirement, and other factors like humidity and air flow rates may need to be considered. The functional needs of industrial buildings often impose severe limitations on their heating systems. Many have high roofs, necessary for plant or machinery, and they are often adapted from previous uses. Exposed bays, open doors, intermittent occupancy and low heat capacity all add their constraints. Until all these factors and requirements *are* determined, there can be little hope of economic operation.

For control purposes, the energy used for space heating over a fixed period of time should be related in some way to the prevailing weather conditions. A useful way of doing this is to compare energy consumption with degree days. Degree days are equivalent to the number of degrees (°C or °F) by which the mean outside temperature over a 24-hour period lies below a predetermined base (15.5°C, 60°F). Thus a 24-hour period in which the mean outside temperature is 12.5°C will have 3 degree days (°C). Temperatures above the base are excluded. The total number of degree days over a period gives a numerical indication of the average 'coldness' for the locality and hence the amount of heat which should be necessary for space heating. Degree day data based on regular temperature recording by meteorological stations are published by various organisations, for example, in the UK, by the Department of Energy. The method is approximate only and reasonable allowances have to be made for such things as hot water use and incidental heat gains. Despite its limitations, however, 'energy used per degree day' can be a useful practical tool for comparing consumptions provided the time periods selected are not too short. Monthly comparisons are generally most satisfactory.

Whatever levels of heating, cooling, humidification or air flow are deemed necessary, the objective is generally to produce a steady, balanced state regardless of outside conditions. Because of climatic variations in most parts of the world, however, this is often difficult to achieve economically. *Perfection is expensive*, and the amount of compromise possible in order to minimise operating costs deserves careful consideration.

The human body reacts quickly to environmental change. Not only is it sensitive to the average dry-bulb temperature, it is also sensitive to radiation, air flow, the relative humidity of the air and to temperature differences between floor and head height. An average air temperature of 18–21°C (65–70°) is sufficient for most work involving little physical activity, but this depends on other factors being satisfactory. Floor temperatures much below this level will cause discomfort, as will relative humidity outside the range 40–70%. Lower air temperatures may be acceptable if some of the heating is from a radiant source, or if air velocities are low, below 0.2 m/s. Walls that are below the air temperature tend to cause discomfort to those seated near them.

An air change of one per hour is sufficient in many working situations to prevent air becoming stale, but in industry, air change rates well above this are common. These may be uncontrolled (e.g. through doors, windows and loading bays), or they may be essential in order to remove fumes, odours, etc. arising from production processes. Heat losses arising from air changes are considered in Section 6.4.

In determining heating requirements, account should be taken of the type of clothing worn by the occupants and also heat gained from process heating, lighting, electric motors, etc., and not least, the sun. Use should be made of these adventitious gains whenever feasible. This is not always easy and large glazed areas, in particular, often result in overheating in summer and high heat losses in winter because of the poor insulating properties of glass, though improvements are possible using solar control glasses and films. In many cases summer working conditions and productivity as well as heat retention in winter will be improved by extra roof insulation.

A proper balance of heating, lighting, acoustic and other requirements can best be achieved in a new building. In recent years some useful studies have been carried out into the concept of fully integrated design, which takes account of these important environmental and economic aspects as well as structural design and aesthetic appeal.

6.2 Heating methods

In selecting the most appropriate heating method, essential if operating economy is to follow, consideration should be given not only to the building design but also to the activities, processes and movements that take place within it.

Convection heating

This uses the circulation of water, steam or, occasionally, heat transfer fluids. Plain panel heaters working by natural convection (known in the UK as 'radiators', although only 20% or so of their output may be radiant) are suitable for buildings with a multiplicity of rooms requiring individual heat control. Such control can readily be effected by the use of simple individual radiator thermostats, though there is perhaps little point in adding these refinements if room occupants are constantly tampering with the settings. Natural convection heating is quiet and suitable also for moderately dusty and flammable atmospheres. It is *not* well suited to buildings with high ceilings or roofs.

Convectors with extended heating surfaces (e.g. finned tubes) are more compact and flexible, particularly when fan-assisted and fitted with adequate controls. Industrial unit heaters of this type, supplied with hot water or steam from a central point and mounted at high level, release valuable floor space, but distribution losses are, of course, higher and each unit needs to be carefully sited and adjusted to give uniform heating at ground level. Such units are best operated at high temperature, that is, using steam or high-pressure hot water, and with on–off thermostatic control.

Air heating

Air heaters, individually fired by distillate oils or gas, are commonly used in industrial buildings. They have the same advantages as unit heaters using steam or high-pressure hot water, and can be used when these services are not available. Because they have no associated distribution losses, they operate at high overall thermal efficiency and have rapid start-up and response, but for large open areas a number of smaller heaters or extended ductwork may be required in order to achieve uniform heating. Units may be free-standing or mounted at high level and warm air can readily be recirculated if required.

Direct-fired ('flueless') air heaters are cheap and useful where air change rates are high or the conditions particularly rugged. For these heaters only distillate oils, liquefied petroleum gases or natural gas should be used and *not* residual oil fuels or waste oil.

Radiant heating

Radiant heaters are useful in high buildings, for localised 'spot' heating and in places where high rates of air change cannot be avoided, e.g. open loading bays. They assist quick start-up on cold mornings and provide a generally more 'comfortable' working environment than do convection or air heaters alone. Panels or tubes mounted above working level may use high-temperature air, steam or high-pressure hot water or may be gas fired. Electric heaters are sometimes used where electric cable is easier to install than pipework. Gas-fired and electric radiant heaters must *not* be used if there is any possible fire hazard, and all forms of radiant heating should be used with care.

The effectiveness of radiant heaters depends upon maintaining high surface temperatures. Reducing hot-water temperatures in overhead panels and tubes,

6/1 A 15 kW (52 000 Btu/h) output overhead gas-fired radiant heater. Products of combustion from the burner are drawn through the radiant steel tube assembly by an induced-draught fan. This provides combustion air at the burner and also air for waste gas dilution so that the heater can be used unflued in many instances. Combustion air and waste gases can readily be ducted, however, if required. Controls are housed in an enclosed box at the burner end. *Hamworthy Enginering Ltd, Poole, Dorset.*

for example, is more likely to *increase* heating costs than to decrease them. If heating capacity is too great, economies should be sought by eliminating some panels or lengths of tube or by using on–off or wider differential control so that, when operating, the panels and tubes can be maintained at high temperature.

Unless they form part of a completely integrated building design, underfloor heating, storage heaters and electric heating are not often used for industrial and commercial space heating because of their higher running costs and lack of flexibility.

In selecting a heating method, too much emphasis should not be placed on quoted full-load thermal efficiencies obtained under laboratory conditions. A unit which is slightly less efficient but which is part of a system that provides more uniform heating at low level and which responds quickly to controls, may give lower overall running costs.

- **Determine the precise purpose for which the heating is needed and the optimum heating and other environmental requirements.**
- **Determine the maximum permissible variation in these requirements in extreme climatic conditions or emergencies.**

- **Ensure that existing and proposed heating methods are suitable for the shape and construction of the buildings and for the processes involved.**
- **In new buldings, consider heating requirements at the outset of planning. Consider the purpose to which each part of the building will be put and challenge the need for large areas of glazing.**
- **Select the size and number of heaters so that each unit can be operated at close to its maximum output.**
- **Where adjacent working areas in the same building have markedly different heating requirements, consider division by partitions or temporary screens.**
- **Do not use more working space than is necessary and provide only minimum heating to prevent freezing in areas not in use.**
- **Maintain all heating plant in good running order using a schedule for regular inspection to ensure:**
  cleanliness and freedom from obstruction and damage of heating surfaces
  cleanliness of air filters, inlets and extracts
  correct lubrication and adjustment of all moving parts
  correct setting of louvres, grilles and dampers.

6.3 Temperature and time control

The actual average temperature of a working environment and the length of time over which this temperature is maintained are vital factors in economical operation. A temperature 2°C higher than necessary can result in fuel wastage of about 10%. A reduction from 8 to 7 hours in the time for which the working temperature is maintained can lead to a saving of about 10%.

Temperature

There are various legal and other minimum temperature requirements (which may be modified by emergency legislation) which must be met, and heating plant should be designed to do this under the worst outside conditions. But plant should not be oversized. Oversized plant operates inefficiently, and estimates of the worst conditions to be met should be *realistic* ones based on local records. They should only *just* be met. This will provide ample margin for all *normal* conditions.

Control of temperature should be by thermostats rather than manually. The correct location of the sensing element of a thermostat is vital. It must be in a representative position for the zone which it serves, and this is best determined by a careful site inspection, rather than from building plans which may not show all the features that must be considered. Never locate the sensing element:

on an outside wall
near a window or door
in direct sunlight
in a hot air stream
within direct sight of any radiant heat source (e.g. a furnace wall).

Areas to be heated should be carefully zoned. Not all zones require the same temperature and each one should have separate thermostatic control. Not too many units should be controlled by the same thermostat. Parts of any building which receive direct heat from the sun should be separately zoned, and special provision made to compensate for the heat received, particularly in spring and autumn. Generally, lower working temperatures are acceptable when temperatures are uniform and draughts are eliminated. Greater thermostat sensitivity enables lower set-points to be used.

Time

Various legal requirements must also be met for time as well as temperature. Workplaces of all sorts must attain a reasonable temperature within a stated period from the start of work. In practice, many buildings attain the specified temperature long before it is necessary because the heating plant is started by automatic time control which operates regardless of the outside temperature. The problem is less acute where quick-response systems like air heaters or radiant heaters are used and worst in buildings with high thermal capacity with heating systems of high thermal inertia (conventional hot water circulation).

Control devices which delay the start-up of heating plant according to the outside air temperature, inside air temperature and thermal characteristics of the building structure are now available and should certainly be considered for any large space heating load. Such optimum-start controllers and more sophisticated building management controls should, of course, be used in *addition* to sound practice in every other aspect of heating. They are not a substitute for it.

Similar economies may be effected at the end of a working day or shift by shutting down heating plant before work ceases,the precise time depending on the heat capacity of the building and heating system. All workplaces are not necessarily in use for the same periods of time and automatic time controls should be set (or overridden) accordingly. The advantages of flexible-time working, employed in some organisations, are offset to some extent by the need to maintain various building services for longer periods each day than would otherwise be necessary.

All controls, and particularly those for temperature and time, should be as simple as possible and easily understood and maintained – the installer's enthusiasm should not be allowed to exceed the user's ability. Controls should be regularly checked and any unauthorised tampering with the settings should be investigated. Locked or tamper-proof controls are an obvious advantage. Unattended boiler rooms which contain controls should be kept locked.

- **Ensure that all heaters are thermostatically controlled and that only units in the same general zone as the thermostat element are controlled by it.**
- **Ensure that all thermostat elements are correctly sited in representative positions.**
- **Determine the optimum temperature for each building or zone and set each thermostat accordingly. Record the temperatures and check them regularly using an accurate thermometer.**
- **Make use of the sun by controlling separately any part of a building which receives a substantial amount of sunlight.**
- **Do not start up heating any earlier than is absolutely necessary before the working period begins.**
- **Switch off heating before the end of the working period. Experiment to find the earliest acceptable cut-off point.**
- **Unless reliable manual control is possible at no extra cost, install automatic controls to start up and switch off heating. For larger buildings, consider installing optimum-start controllers which take account of outside temperature.**
- **Ensure that the heating system is not brought into use (except for frost protection) at weekends, during holidays or during any period when a building is not occupied.**

6.4 Air changes and heat distribution

Unnecessary air changes in industrial buildings are one of the main causes of heat loss. The number of air changes necessary to prevent air becoming stale and humid varies with the number of occupants and the processes carried out within a building. Data on acceptable air change levels are contained in the CIBS and similar guides. Where no accepted standard exists for a particular situation, expert advice should be sought. The *minimum* level is one change of air per hour in a workplace where no fumes or dust are produced and smoking is not permitted. In industry, the problem is often one of reducing unnecessary and uncontrolled air changes rather than increasing the amount of replacement

6/2 Roof ventilation at London's new Covent Garden Market. The units require no electric power and are regulated pneumatically to avoid overventilation and unnecessary heat loss in winter. Badly sited and continuously operated exhaust fans waste energy and modern extraction units like these give much better control of air changes. *Colt International Ltd, Havant, Hampshire.*

fresh air to maintain correct conditions. In most factory buildings one to two changes of air per hour occur by normal infiltration. Each hourly air change is equivalent to a heat loss of 0.33 W/m^3 of building volume for every °C difference between inside and outside temperatures.

Leakage of warm air and infiltration of cold air occur at various points in a building, and also through the incorrect operation of exhaust fans, ventilators and other mechanical equipment. Leakage and infiltration can only be dealt with by a thorough and systematic survey of the fabric of the building to ensure that it is wind-tight, attention being paid to displaced cladding, missing tiles, broken windows and gaps in and around ridges, eaves, windows and doors.

6/3 An ambient air curtain is especially useful where main doors have to be left open for long periods. Air, drawn from high level within the building, is blown into an underground duct from which it is projected through slots at high velocity. This prevents cold air ingress, even when rolling stock is standing in the doorway.
Bahco Ventilation Ltd, Banbury, Oxfordshire.

Industrial draught-proofing services are now available to deal with ill-fitting windows and doors.

Doorways and loading bays present a particular challenge to ingenuity. Warm-air curtains were a popular solution in the days of cheap oil, and there is no reason why they should not still be, provided that they are correctly installed so that *all* the warmed air enters the building and no more is warmed than absolutely necessary. Many such curtains operate only when the doors are open, with the obvious danger that, as a result, they are left open longer than really necessary.

Ambient air curtains are an interesting solution to the problem, especially when doors *do* have to be left open for extended periods. Fans draw warmed air from above the door level and project it across the opening at high velocity to form a positive barrier. For heavier duty (e.g. fabrication shops) *external* ambient air curtains can be used, though drivers are sometimes less than enthusiastic about the cold air barrier which this type provides in winter.

Automatically operated doors have gained in popularity in recent years. Using air-operated thrusters, hinged flexible doors can be opened — and closed — automatically by signals from a variety of sources, ranging from driver-operated push buttons to vehicle detector loops embedded in the floor. Sliding doors can be automated in a similar way and in many organisations this is the preferred method of dealing with the door problem.

Provided that they are installed and used with reasonable care, plastic strip curtains can provide an economical and flexible method of preventing heat losses through doors which are constantly in use. Drivers of fork-lift trucks without enclosed cabs may need guards to protect them from the edges of the strips and personnel on foot should be discouraged from using doorways which are normally used by vehicles. Torn or damaged strips should be replaced promptly. As well as keeping warm air in, plastic strip curtains are useful, too, for retaining cold air in refrigerated areas. Most manufacturers can supply special grades of material for low-temperature applications.

6/4 The automatic opening and closing of external and internal doors is often the most satisfactory solution to heat losses and unwanted air movements. A variety of sensing devices is available. In this bonded warehouse, automatic sliding doors, with transparent panels to improve safety, are fitted behind the folding steel security doors. They maximise storage space and minimise damage to the loads carried.
Harefield Rubber Co. Ltd, Harefield, Middlesex UB9 6HG.

6/5 Heat losses through doorways in frequent use can be drastically cut by the use of suspended transparent strip curtains. Such curtains are also useful for internal doorways to minimise draughts and reduce air movements between differently heated zones. For low-temperature applications, special grades of material are available.
Harefield Rubber Co. Ltd, Harefield, Middlesex UB9 6HG.

If loading bays can be isolated from other working areas and especially if through-draughts can be prevented, conditions can be improved and heat saved. When bays have to be left open to the elements, local overhead radiant heaters should be used. Whenever possible, vehicles being loaded and unloaded should be completely enclosed within the loading area or, if they are of reasonably uniform size and rear-loaded, backed up to purpose-built flexible dock shelters. Nearly all large doors for vehicles benefit from the addition of a small self-closing personnel door alongside or incorporated in them.

Fans and powered vents of all sorts are potential sources of heat loss. Large extract units may provide opportunities for heat recovery, especially in situations — such as filling plants for flammable materials — where high air change rates are essential. All too often, though, they lead only to overventilation and consequent heat losses. The use of all powered air extraction should be challenged and superfluous units should be disconnected, especially in winter.

Recirculation of warm air at roof level back down to working level, however, is a form of 'ventilation' which reduces heat losses and which, therefore, has much to commend it. Such recirculation can be by means of fans and ductwork, possibly to the air inlets of floor-mounted air heaters, or by self-contained high-level units which propel warm air downwards against its natural inclination to provide more uniform temperatures. Enclosed fans and large open propeller fans are available.

6/6 Warm air at roof level returned to working level by means of a recirculation unit consisting of a small low-power fan and a lightweight flexible circular duct. By reducing heat stratification in buildings, working conditions are improved at no additional cost for heating and heat losses through the roof are lowered. *Dohm London Ltd, London SE19 1PL*

Overventilation is common where particular processes produce fumes, oil mists and smoke which, for health reasons, must be removed or dispersed. Doors opened wide are a common sight in welding shops. In many instances, local fume collection units can be used, and these not only provide a high degree of health protection to operators, but also return the cleaned warm air to the building. Even where large volumes of air are contaminated, a high proportion of the air extracted can often be recycled after being cleaned and deodorised by passing through activated carbon filters. The cost of this sophisticated treatment is still far less than the cost of heating cold replacement air. When there really is no alternative to extracting and completely discarding air from particularly noxious processes, care should be taken to ensure that the extraction unit is not so oversized or badly positioned that it also removes clean warm air from the building along with the foul air. If possible, such process areas should be segregated and dealt with separately from areas where the air is comparatively clean.

- **Determine the optimum number of air changes per hour required in each building or zone, taking account of the nature of any process carried out.**

- **Examine the fabric of buildings and eliminate all uncontrolled air flow. Ensure that roofs, walls and windows are sound and minimise air flow in cavities.**
- **Search for and eliminate losses through old chimneys and ducts. Be certain that they are no longer used before blocking them off.**
- **Ensure that all doors and windows fit correctly and are firmly closed whenever the heating is on. Exterior doors should be self-closing.**
- **Check evenness of heat distribution by regular measurement at selected points and ensure that there are no obstructions to warm air flow.**
- **Pay particular attention to loading bays and to adjacent workplaces.**
- **Isolate, as far as possible, noxious processes and provide local extraction equipment, positioned to maximise the extraction of contaminant, and to minimise the extraction of clean warm air.**
- **Avoid mixtures of natural and powered ventilation, and check that ventilators required for summer use are shut in winter.**
- **If general ventilation rates are excessive, eliminate some exhaust fans. Complaints of cold draughts are likely to be reduced if the building is maintained at a slightly positive pressure by a suitable balance of input and exhaust fans.**
- **To minimise heat stratification and high losses at roof level, recirculate clean warm air to low level.**
- **Check the ceiling height actually required in all rooms and buildings. If grossly excessive, consider the use of false ceilings.**

6.5 Building insulation

U values

Thermal transmittance coefficients or *U* values indicate the heat transmitted per unit area for each degree of temperature difference, i.e. watts per square metre per °C or Btu per hour per square foot per °F. (To convert Btu/ft^2 h °F to W/m^2 °C multiply by 5.68.) The heat loss rate through a structure is proportional to its *U* value and the total heat loss is the product of total area, temperature difference and *U* value.

A *U* value relates to the structure as a whole, and is the reciprocal of the sum of all the individual resistances. It takes account of surface resistances and resistances of cavities, air gaps, etc., as well as the thickness and thermal characteristics of each component in a composite structure. A U value is *meaningless* unless the components, thicknesses, construction and conditions of use are properly specified. Typical values are listed in many reference books and in manufacturers' literature. A precise value should be obtained for any proposed structure or modification.

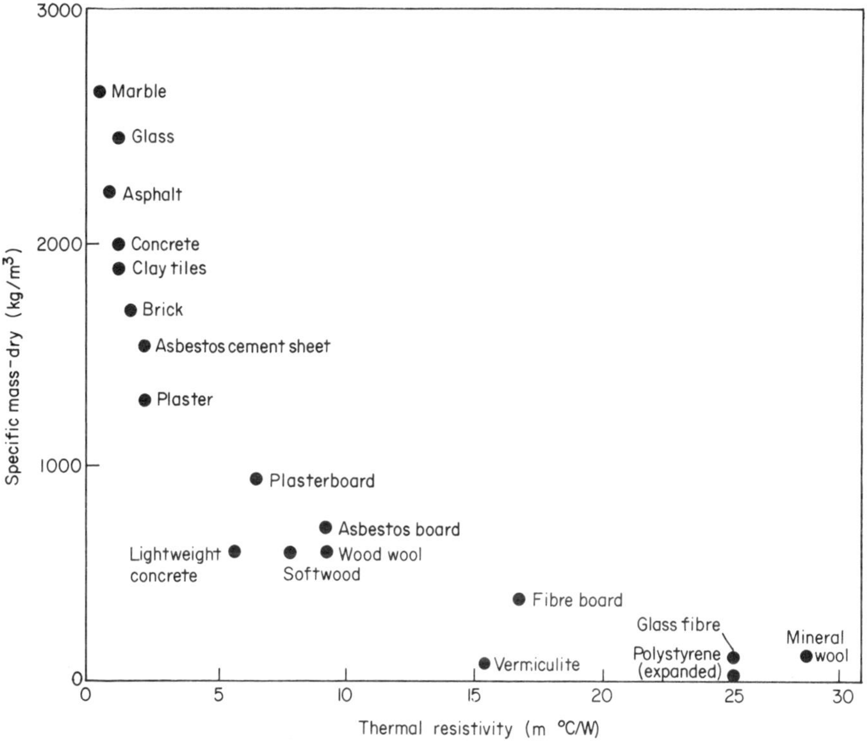

6/7 The thermal resistivity and bulk density of some common building construction and insulating materials. Exposure to the weather may result in lower thermal resistivity values (e.g. for brick and concrete, about 25–30% lower).

Resistance to heat flow is greatest in materials which contain a high percentage of small air pockets or cells, and heat flow through all building structures depends on air layers and cavities as well as on the properties of the materials themselves. For this reason, U values (thermal transmittance, W/m^2°C) should always be accompanied by an accurate description of the material or structure. Values for a wide range of materials can be found in section A3 of the *CIBS Building Services Manual.* (Based on data compiled by P. L. Martin.)

A wide variety of insulating materials is available, some of which also have structural or load-bearing properties, or incorporate decorative finishes. In selecting insulation a number of factors should be carefully considered:

Thermal properties at the temperature of use.
Mechanical properties and susceptibility to mechanical damage.
Fire resistance, and smoke and fume production in the event of fire.
General resistance to water, chemicals, oil, fungi, etc.
Likelihood of infestation by vermin.
Vapour barrier properties.
Price.

Legal limits to heat losses from buildings should not be used as the sole criterion for the selection of thermal insulation. Such limits should be *more* than met and after the factors listed above have been considered, the most cost-effective type and thickness of insulation should be determined by economic evaluation, using an estimated future cost of heat at the point of use.

The order of priority for insulation of an industrial building will normally be as follows.

Roof. Foamed, expanded or fibrous material should be placed between roof joists or between existing lining and cladding. For old roofs with no insulation or lining, rigid material may be attached to the underside of the cladding or in suitable forms laid on top of the existing roof structure.

Walls. Cavities can be injected with suitable foamed, expanded or fibrous material. This should only be undertaken if a proper, independent survey indicates that the wall structure is suitable for such treatment. Solid or single-skin walls may be lined internally with a lightweight insulating block wall or with rigid insulation. Rigid slabs of suitable material may also be incorporated in the cavity walls of new buildings. Insulation of walls adjacent to heaters is important. It has been shown, for example, that for buildings heated by radiators on external walls, the simple application of reflecting material — aluminium foil — behind the radiators yields savings of 5% for solid 230 mm (9 in) brick walls and over 3% for 280 mm (11 in) cavity walls.

Windows. Broken windows should be replaced with heavyweight glass or by double glazing. Lining with plastic sheet may be possible if fire or other safety regulations are not infringed. Windows present a difficult problem as far as heat losses are concerned. Careful consideration should be given to the purpose of and need for all windows, and complete covering-in or bricking-up should be undertaken if a window is redundant. The value of natural light from rooflights needs to be carefully assessed and weighed against the cost of heat losses.

Floors. Lining, tiling or carpeting will improve the thermal properties.

With an improvement in the insulating properties of any structure, great care must be taken to ensure that, as a result, condensation does not then take place on the outer surfaces made cooler by the improvement. Suitable barriers must be used to prevent the migration of water vapour. Many aspects of thermal insulation may have secondary effects on the building structure and the guidance of a qualified building surveyor should be sought before any major modification is undertaken.

- **Assess the thermal characteristics of industrial buildings in the following order of importance: roof, walls, windows, floor. Consider the feasibility of reducing heat losses by the application of *suitable* insulating materials.**
- **When sound practical improvements seem possible, carry out a careful assessment of the financial savings which will be made, using the estimated future cost of *heat* at the point of use (and *not* the current delivered cost of fuel).**
- **Ensure that all insulation is entirely suitable for the service intended and that it will not be vulnerable to mechanical damage, moisture or vapour ingress, infestation or deterioration.**
- **Ensure that all modification to reduce heat losses comply with current fire, safety and building regulations and codes.**
- **Ensure that, in the event of fire, insulation will not increase the hazard, or result in the dangerous generation of smoke or fumes.**
- **Ensure that the application of insulation will not cause a condensation problem. Provide adequate vapour barriers.**

7 FURNACES AND DRIERS

7.1 Thermal efficiency

The terms 'furnaces', 'kilns', 'driers' and 'stoves' are applied to a whole range of fuel-consuming plant other than boilers. The variety of types and purposes is very wide and each requires individual consideration. Some requirements for economic operation are fairly common, however, and they are considered in this section, using the term 'furnaces' to cover them all. Some special aspects of industrial drying are also considered.

In all furnaces, waste gases must obviously be discharged from the heating zone at temperatures above those to which the stock is heated. In addition, in order to complete chemical or physical changes the stock may need to be held at constant temperature for some time. In high-temperature processes, particularly, high thermal efficiency is clearly impossible within the furnace itself and it is attained only by heat recovery *outside* the structure, for example by preheating combustion air or by load recuperation (*see* Chapter 8).

Efficiency in furnaces is not always easy to define. It varies during process cycles and during 'soaking' periods it may be zero. An indication of the energy required to do specific work — energy productivity or index of performance, e.g. GJ/tonne processed — is often more meaningful than an assessment of actual average thermal efficiency.

7.2 Furnace structure, firing and controls

Furnaces should be free from leaks and damage. Cracks in brickwork and hot-air ducts, doors and dampers that fit badly, damaged refractories and insulation are all likely sources of heat loss that should be attended to. The infiltration of cold air can often be reduced substantially by adjusting dampers to operate the furnace at a very slightly positive pressure, provided that this can be tolerated in the general working environment.

To maintain correct temperatures and furnace atmosphere conditions — oxidising or reducing — throughout the operating range, the firing and control systems should be carefully selected and properly serviced. There are a number of variables which affect the close control of combustion air and when the air is preheated this may not be easy. In most furnaces waste gas temperatures are predetermined and *cannot* be reduced without affecting performance so that air control is very important for economic operation. However, excess air levels should not be reduced so far — particularly with gas firing — that carbon monoxide is produced in quantities above about 1000 ppm as this will reduce

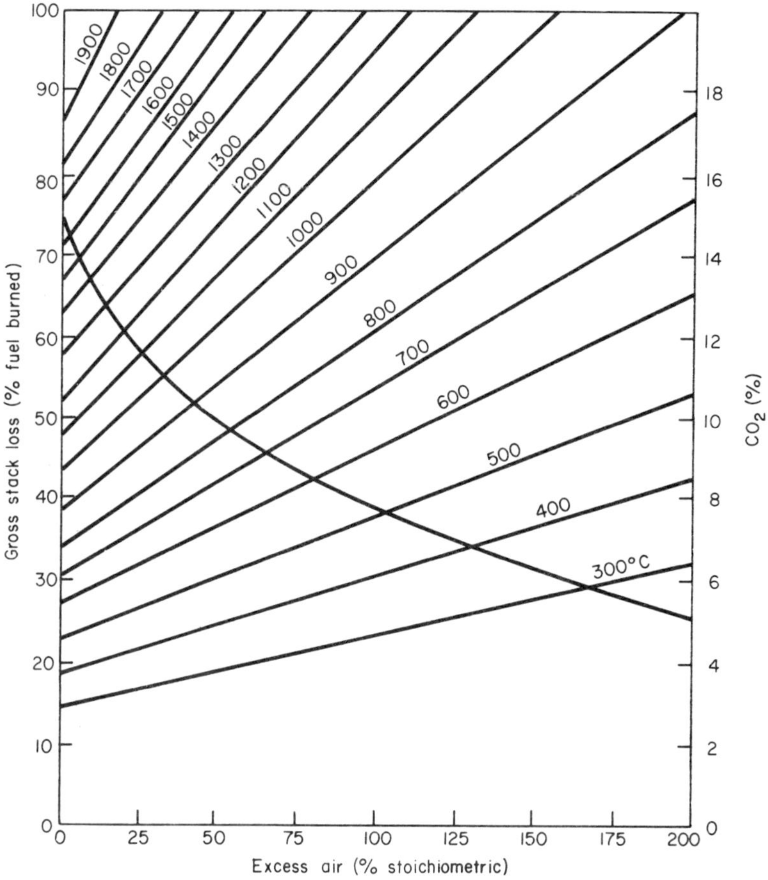

7/1 Waste gas heat losses for oil-fired furnaces at various temperatures (°C). As the temperatures of waste gases increase, heat losses become increasingly significant even at low excess air levels. For high-temperature furnaces, heat recovery is essential. *Shell UK Oil, London WC2R 0DX.*

overall efficiency owing to the loss of potential heat in the carbon monoxide.

Firing equipment should produce flames of the correct shape and size. Flame impingement on the furnace lining must be avoided as must flame 'sting' through openings and doors. Combustion in furnaces is generally much more tolerant than in boilers, but correct fuel preparation, constant supply pressure, etc., are still needed if firing equipment and controls are to function correctly. Uniform fuel supply conditions are vital if maximum and minimum firing rates are to be accurately set before adjusting air supply rates.

- **Judge performance by comparing the heat input required to do a specific amount of heating or drying under standard conditions.**
- **Ensure that furnace structures and insulation are in good order and that doors and dampers fit correctly.**
- **If possible, set dampers to give a slightly positive pressure in the furnace to prevent cold-air infiltration.**
- **Avoid fluctuations in fuel and air supplies to the firing equipment.**
- **Select firing and control equipment to give the furnace atmosphere conditions required. Minimise excess air, but do not produce carbon monoxide if a reducing atmosphere is not needed.**
- **Avoid flame impingement and 'sting'. Aim for a uniform flame which just fills the combustion space intended for it.**

7.3 Process methods and materials

The most appropriate method of heating should be reviewed periodically. New lightweight furnace structures and more sophisticated refractory and insulating materials give scope to designers and operators to improve performance. New heating methods and firing equipment are continually being developed. Larger furnaces give promise of greater efficiency but only, of course, if they are fully loaded, and, as with all new developments, the temptation to change should be resisted if the promised advantages cannot be fully realised by effective utilisation.

The precise purpose of all driers and stoves, particularly, should be challenged. The extent to which materials need to be dried and the possibilities of reducing moisture contents by mechanical means — centrifuging, filtering, squeezing, etc. — should be investigated. For metal finishing, equivalent alternative materials which can be applied at lower temperatures and with better overall economy should be sought.

Direct firing using premium fuels instead of indirect (muffle) firing may be possible. The sensitivity of the material being processed (especially foodstuffs) to direct contact with the waste gases should be investigated first. Improved efficiency from direct firing may more than offset the additional cost of a more refined fuel and justify the extra effort needed to maintain perfect combustion and avoid contamination.

When fuel was comparatively cheap there was a tendency to 'buy time' and to overfire furnaces in order to maximise production rates. This is now very expensive. Rounded values, like 400°C for 4 hours, should be regarded with the utmost suspicion and precise heating cycles should be determined.

The reduction of processing losses of all sorts, including oxidation, can make a useful contribution to cutting energy costs. Materials which have to be reprocessed — or worse still, scrapped — waste energy. What may at first sight appear to be an expensive heating method may in fact be cheaper if losses are reduced. This applies particularly to the use of low-sulphur-content premium fuels like LPG and natural gas and to various forms of electric heating. The efficiency of electricity generation is low and it is generally expensive compared with fossil fuels, but using electricity can lead to reduced losses, lower overall energy consumption and lower costs to the user. A number of electro-heating methods are available including direct resistance and induction heating, dielectric loss and microwave absorption as well as more conventional indirect resistance heating.

- **Review the precise purpose of all furnaces and ensure that each is entirely suited to its use.**
- **Consider whether or not a change in the process method or the materials used would reduce the amount of energy required.**
- **Consider direct firing using premium fuels instead of muffle firing.**
- **Challenge heating cycles. Be suspicious of rounded values. Use high temperatures where possible but do not overfire in order to save time.**
- **Reprocessing and scrapping squander energy. Investigate heating methods which will reduce such losses even if at first sight they appear more expensive.**

7.4 Heat transfer

In most fuel-fired furnaces heat transfer is by radiation or convection, or a combination of these. For every furnace the main heat transfer requirements should be reviewed. This may well indicate ways of improving efficiency. In all cases, however, the maximum rate of heat transfer that can be tolerated by the material being heated and the maximum operating temperature of the refractories used must be considered *first.*

Radiation

Radiant heat transfer depends on the difference between the fourth power of the absolute temperature of the emitter and that of the receiver. There is therefore a great advantage in using the highest possible flame and refractory temperatures. This is usually achieved by minimising excess air and preheating combustion air.

Heat transfer also depends upon the relative emissivities, distance apart and effective areas of both emitter and receiver. In general, the transfer rate is increased by the use of a flame of high emissivity, by minimising the distance between the flame (or refractory) and the material being heated and by having both of similar area. Emissivity is usually expressed relative to that of a 'black body'. A perfect black body absorbs all radiation, reflecting and transmitting none. Flame emissivity is highest when it contains carbon particles and also when the flame is thick, though thickness may need sometimes to be sacrificed in order to obtain good flame coverage. 'Dead-burned' gases also contribute to

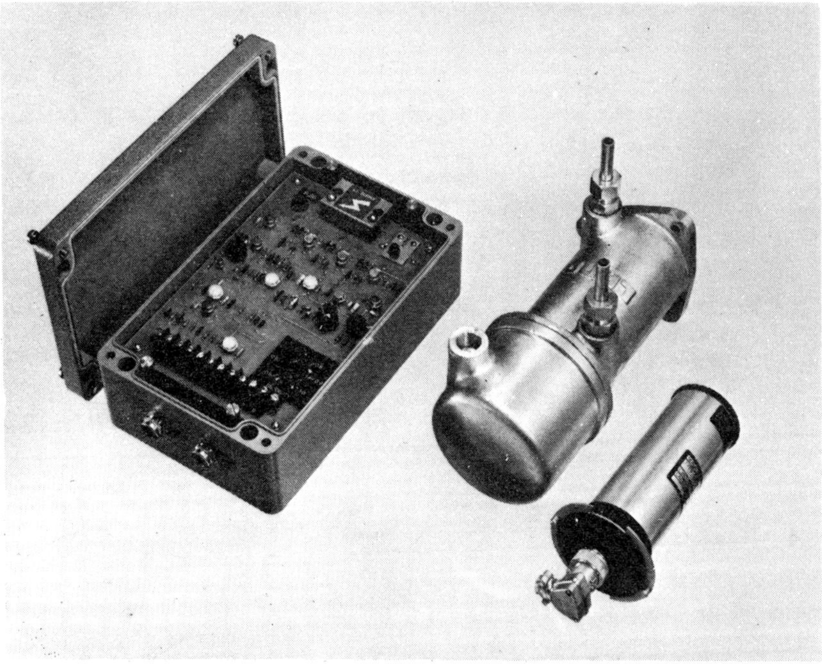

7/2 Remote reading of temperatures is required in many industrial situations, not only for high-temperature furnaces but also for inaccessible spots and moving objects and surfaces. A radiation pyrometer (thermometer) measures temperature by focusing radiation (which may be non-visible) from the source onto a thermopile detector, which generates an emf.

Components of an infra-red thermometer system are illustrated. The thermometer (right) is housed in a water-cooled protection jacket (centre) from which it can readily be removed. Proper protection and cooling, as well as cleanliness of the lens, are essential for trouble-free operation. The electronic signal processing unit (left) can handle a range of functions such as emissivity correction, intermittent or varying signal compensation, etc

Land Pyrometers Ltd, Sheffield, South Yorkshire.

radiant heat transfer, their 'non-luminous' contribution depending on their carbon dioxide and water content as well as their temperature.

Convection

Convection heat transfer rates are approximately proportional to the power 1.25 of the temperature difference between the hot gases and the material being heated so that high gas temperature is important. Equally important, however, are the area across which heat exchange occurs, the thermal resistance of any film of stagnant gas on the surface of the material being heated and also the mass flow of hot gases. Convective heat transfer is therefore promoted by turbulent flow, clean heat exchange surfaces, high velocities and high temperature differences.

- **Determine the principal method of heat transfer in each furnace. Optimise conditions to give maximum heat transfer rates for the method used.**
- **For radiant heat transfer:**
  keep flame and refractory temperatures high
  ensure that the flame is highly luminous
  reduce the distance between flame and stock if possible
  maintain good flame coverage and load the hearth area fully.
- **For convective heat transfer:**
  keep gas temperatures high
  allow adequate circulation space and eliminate stagnant pockets
  maintain high gas velocities and promote recirculation
  use waste gases for dilution rather than cold air
  keep surfaces clean and avoid stagnant surface layers.

7.5 Operating procedures

Sound operating procedures probably yield the greatest energy savings of all, particularly for high-temperature furnaces. A detailed study of process flow is vital in order to determine optimum procedures and to match furnace outputs to other processes. When there is a choice, the more efficient furnaces should be used and they should be loaded to capacity. The possibility of improving the *effective* capacities of furnaces by minor modifications (e.g. lowering the hearth of a bogie furnace) should be investigated.

Frequent intermittent operation should be avoided and every opportunity sought for long continuous operation at high loadings, even if this means some stockpiling. Continuous working for several days instead of single- or two-shift working should be considered. When intermittent operation is unavoidable measures can still be taken to improve overall efficiency. The number and size of furnaces held ready for immediate use should be minimised. For those that *are* held, cooling and recovery rate curves should be determined for each so that, for any known idle period, the optimum cooling and temperature recovery times can be used. In this way a furnace can be ready for use when required without its being held at a temperature higher than absolutely necessary.

Furnace doors should be opened for as short a time as possible and sealed when not in use to minimise heat losses. The heat capacity of carriers and fittings should be as low as possible to improve heating-up rates and to cut heat losses when these items subsequently cool before reloading is carried out.

Where water cooling is used for any part of a furnace (e.g. skid pipes or doors), close attention should be paid to water inlet and outlet temperatures and to flow rates in order to cut heat losses without impairing the effectiveness of the cooling. The use of a thin refractory coating on non-essential parts of the cooling system should be considered.

- **Carry out a detailed study of existing process flow patterns. Modify furnaces to improve effective capacity if possible.**
- **When a choice is available, always use the most efficient furnaces and load them fully.**
- **Avoid intermittent operation. Long continuous operation is much more efficient. Stockpile if necessary.**
- **Determine cooling and heating rates and use optimum cooling and temperature recovery times when furnaces are to be left idle for known periods.**
- **Open furnace doors for minimum periods when loading and unloading. Seal up idle furnaces.**
- **Use lightweight carriers and fittings and no more than necessary.**
- **Minimise the use of cooling water. Reduce heat pick-up within the furnace if it serves no useful purpose by insulating pipes, skids, etc.**

7.6 Refractories and insulation

Transmission through the walls is not usually the major source of heat loss in fuel-fired furnaces. Most furnaces are reasonably insulated — they have to be in order to attain and maintain their temperatures. But savings can still be made, especially in high-temperature and intermittent plant, by attention to insulation.

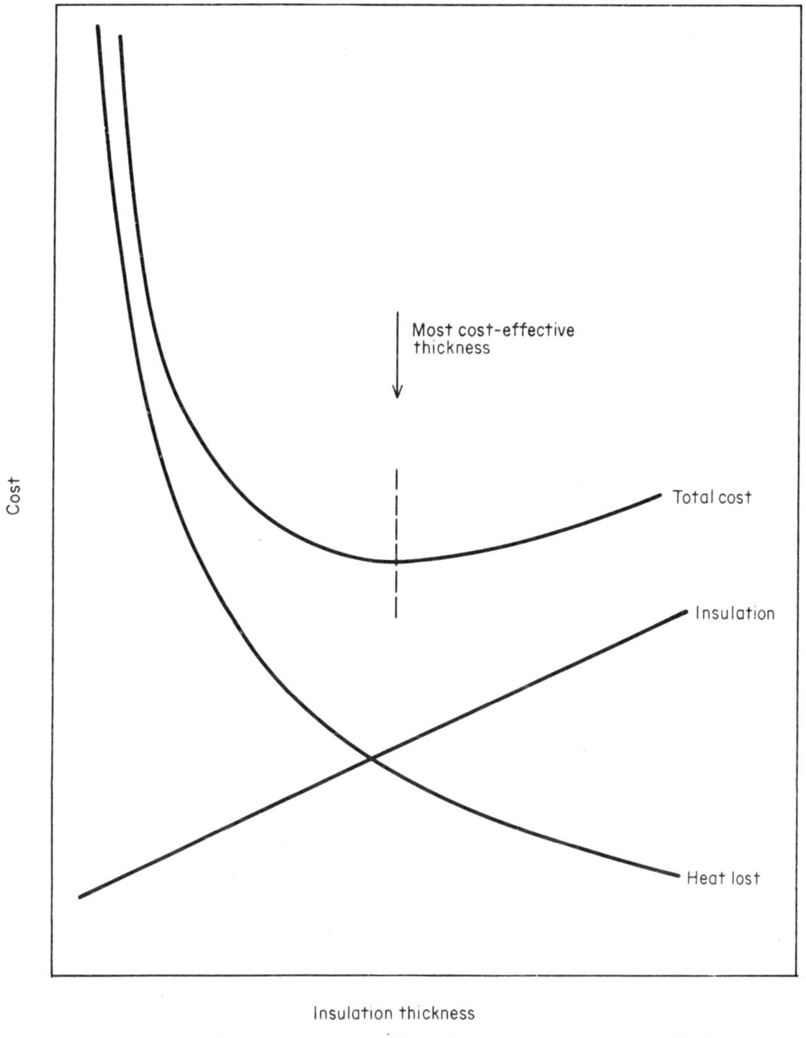

7/3 The most cost-effective thickness of insulation or combination of refractory and insulating materials is that which gives the lowest total of the cost of lost heat plus the cost of insulation. This is illustrated for a single layer of insulation. In practice the costs are usually tabulated for various thicknesses or combinations of materials.

The cost of lost heat should include a share of the variable costs associated with the heat-producing plant, as well as the actual fuel cost, allowing for future increases. Similarly, insulation cost should include installation and maintenance costs as well as the cost of materials. Recommendations are given in BS 5422:1977 and in various charts and tables produced by insulation manufacturers.

Changes to existing furnaces should be undertaken with great care. Exterior insulation should not simply be added without regard to the consequences. A reduction in heat loss through a wall will result in higher internal face and interface temperatures and it is vital that the materials used should still be suitable. The heat loss rate *may* have been carefully calculated to prolong refractory life.

Where an overall improvement seems possible, however, the most cost-effective combination of refractory and insulating materials needs to be selected. The *precise* duty required of the hot-face refractory usually allows no compromise in its selection, but various combinations of backing insulation may be possible and the best overall structure can only be determined by careful expert analysis.

Where intermittent heating is unavoidable, it is particularly important that the heat capacity of the furnace structure should be as low as possible. This can be achieved in some cases by the use of hot-face refractories with good insulating properties but such lightweight materials may not always be suitable for the rough usage given to many furnaces. Recent developments in the use of refractory fibres of low thermal conductivity and heat capacity has greatly extended the techniques available for low thermal mass construction. Service temperature limits have been raised progressively for various refractory fibres and in particular 'ceramic fibres' based on alumina and silica, with chromium and other oxide additions. Ceramic fibres are now available with a continuous-use temperature of about 1400°C. As well as their use in new furnaces, resilient ceramic fibre blanket modules can be used to line existing furnaces, extending their useful life and improving their performance. Lower-temperature batch-type furnaces can benefit considerably from this type of construction which, because it is much lighter, requires less structural steelwork.

- **Do *not* improve the insulation of any high-temperature furnace without first assessing the possible side-effects.**
- **Determine the most cost-effective combination of refractory and insulation, using realistic costs.**
- **Whenever possible reduce the heat capacity of intermittent furnaces by the use of insulating refractories for the hot face, or by the use of ceramic fibre linings.**

7.7 Drying

Drying, the removal of moisture from solids, is a specialised field in which the physical and chemical properties of the materials being dried are of paramount importance. There is considerable variety in drying methods and some very sophisticated equipment is available. Only the basic principles are considered here.

Mechanical alternatives to thermal drying, referred to in Section 7.3, should be reviewed periodically and close attention should be paid to the maintenance of equipment installed to remove water before drying is undertaken. The viscosity of water decreases as its temperature is raised and the efficiency of de-watering processes can sometimes be improved by preheating the material to be dried, for example by the use of flash steam from the heater batteries of the drier. Natural drying should not be ignored and scope sometimes exists for this, such as by drawing material from the surface rather than the inside of a stock-pile.

Drying normally takes place in three fairly well defined phases:

The *heating-up period*, during which the temperature of the material and the water is raised.

The *constant-rate period*, during which the surface of the material being dried behaves as a sheet of water — continuously replenished as evaporation takes place at a uniform rate.

The *falling-rate period*, during which the water is unable to diffuse to the surface at a fast enough rate to maintain it in a saturated condition.

During the constant-rate period, the drying rate is determined by the temperature, humidity and velocity of the drying air, and is limited only by the rate at which water can be taken up by the air.

During the falling-rate period, the plane of drying recedes into the material and the drying rate progressively falls. Drying is limited by the rate at which water can diffuse to the surface and by the surface temperature that the material can tolerate.

It is important to determine, for the particular material being dried, when the constant-rate period ends, the point at which the material is said to have a *critical moisture content*. Some typical values are given below.

A useful technique developed by the Sira Institute may be used to check the moisture content of a specific material during drying, by measuring the difference between its temperature and the wet-bulb temperature of the drying air. The measured difference can be used as a process control parameter.

Overdrying is particularly wasteful of energy and of drier capacity. Unless there is some good reason for so doing, materials that will subsequently be stored in contact with the atmosphere should not be dried below their *equilibrium (or regain) moisture contents*, otherwise they will simply reabsorb moisture. Equilibrium moisture contents depend upon the temperature and relative humidity of the atmosphere. As examples, some typical values are given below, together with corresponding critical moisture contents.

	Critical moisture content (% dry weight)	*Equilibrium moisture content* (% dry weight; air at 25°C, 50% RH)
Leather	90–125	16
Newsprint	60–70	5
Soap	20	10
Wood	40–130	9
Wool fabric	30	13

For uniform materials (free from electrical conductors such as salt and carbon) in which the critical moisture content is well above the final desired moisture level, electrical heating methods — radiofrequency (dielectric) and microwave heating — by which the internal temperature can be raised without affecting the surface, are of particular interest.

Overdrying often results from lack of uniformity of moisture within the material being dried and particular attention should be paid to conditions at the drier inlet. A useful application of dielectric heating is profile drying of the web

7/4 The relationship between the principal properties of an air and water-vapour mixture can be shown on a psychrometric chart and a particular condition may be defined by reference to any two of the following:

Dry bulb temperature (DBT, °C)
Wet bulb temperature (WBT, °C)
Relative humidity (%)
Moisture content (kg/kg dry air)
Specific enthalpy or total heat (kJ/kg dry air)
Specific volume (m^3/kg dry air)
Dewpoint (°C)

Wet bulb and dewpoint temperatures are usually indicated on the saturation line (100% RH) and specific enthalpy by means of two outer scales to facilitate the construction of a line through any point of intersection on the chart. Various charts covering different temperature ranges are available and care should be taken always to identify the correct grid lines and scales, especially for wet bulb temperature and specific enthalpy, the grids for which sometimes coincide.

The example shows the properties of an air and water-vapour mixture with a DBT of 64°C and WBT of 43°C.

Copies of the chart may be obtained from the Chartered Institution of Building Services, 49 Cadogan Square, London SW1X 0JB. Definitions are contained in Section C1 of their *Building Services Manual.*

Psychrometric charts for other barometric pressures (in the range 700–1100 mbar) may be obtained from Troup Publications, 30 Old Burlington Street, London W1X 2AE.

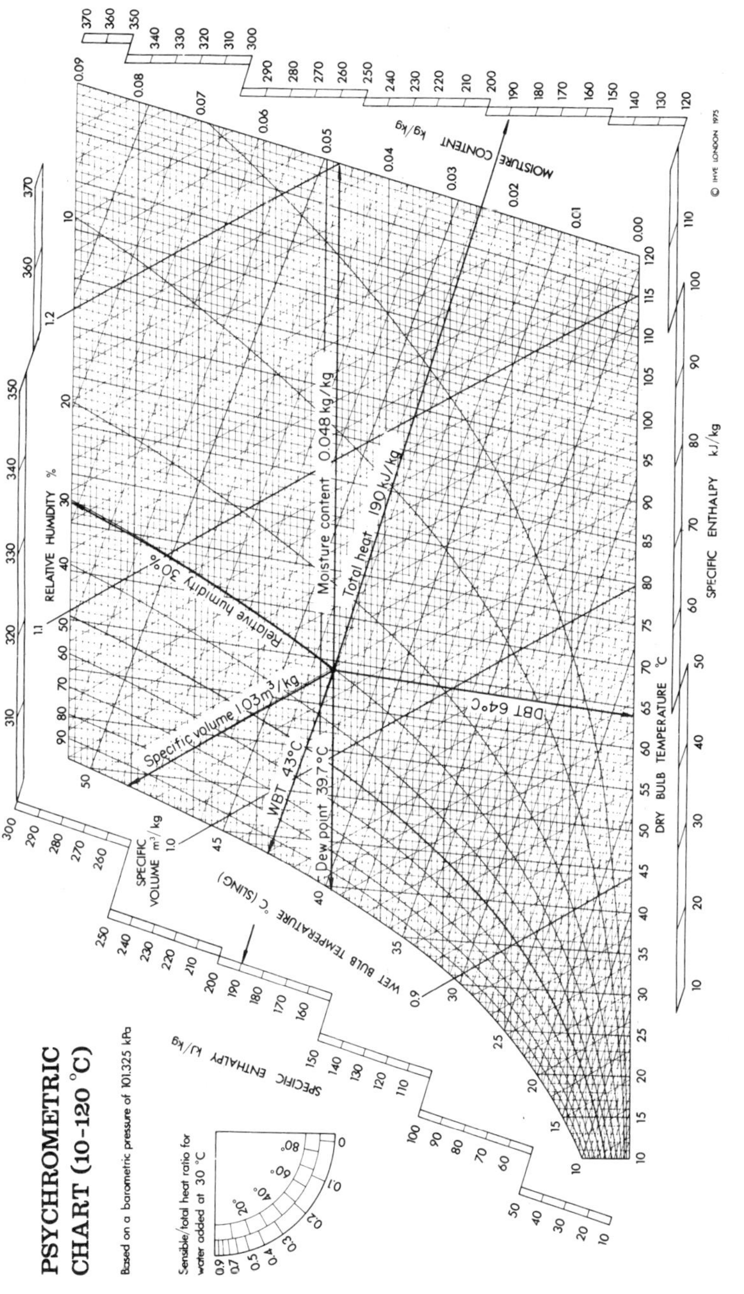
PSYCHROMETRIC CHART (10-120 °C)
Based on a barometric pressure of 101.325 kPa
Sensible/total heat ratio for water added at 30 °C
SPECIFIC ENTHALPY kJ/kg
SPECIFIC VOLUME m³/kg
WET BULB TEMPERATURE °C (SLING)
DRY BULB TEMPERATURE °C
RELATIVE HUMIDITY %
MOISTURE CONTENT kg/kg
Specific volume 1.03 m³/kg
Relative humidity 30%
Moisture content 0.048 kg/kg
Total heat 190 kJ/kg
WBT 43°C
Dew point 39.7°C
DBT 64°C
© IHVE LONDON 1975

on paper-making machines where an uneven moisture profile can be corrected by the self-regulating absorption of high-frequency energy by areas with high moisture contents.

In order to control the temperature of drying air, and, at the same time, increase velocity and decrease relative humidity, combustion gases are sometimes diluted with cold fresh air. This can be a wasteful practice, especially if the fresh air intake is excessive. Study of a standard psychrometric chart will show that as the temperature of air is raised, its potential for holding water increases markedly. The highest temperatures that the material being dried can safely tolerate should therefore be used even if this means operation at higher relative humidity. This can be achieved, and some heat recovered at the same time, by dilution with partially recirculated exhaust air. A relative humidity of about 80% in the drier exhaust is often acceptable, but for each individual process, optimum conditions should be determined.

The general mechanical condition of all driers and predrying equipment needs frequent checking to ensure that rolls, heater batteries, nozzles, dampers, external insulation, etc. are correctly set, clean and generally in good order. Steam leakage into drying air is *particularly* undesirable. As with all energy-using plant, driers should be operated at as high load factors as possible in order to spread the 'no load' losses, which should be measured periodically. Combustion equipment should be well maintained and direct firing practised, whenever feasible. Clean waste gases from *other* combustion processes, especially high-temperature processes, can sometimes be utilised for drying or preheating purposes.

- **Establish the critical and equilibrium moisture contents of the material to be processed.**
- **Remove as much water as possible before thermal drying begins.**
- **Ensure that moisture in the material to be processed is as uniform as possible.**
- **Do not overdry.**
- **Maintain drier temperatures as high as possible.**
- **Consider partial exhaust recirculation as an alternative to fresh air dilution.**
- **Maintain all driers and predrying equipment in good order. Avoid 'short circuits'.**
- **Be vigilant for steam leakage to the drying air.**

8 FLUE GASES AND HEAT RECOVERY

8.1 Chimneys and flues

The continuous supply of air into and the removal of flue gases from the combustion zone are essential for satisfactory burner operation. Both may be achieved by means of the draught (suction) created by buoyancy within a chimney, the use of induced- or forced-draught fans or any combination of these. Once, the main purpose of chimneys was to create draught, but chimneys now have an additional function, to ensure adequate dispersal of the gases and solids that arise from combustion. In *all* parts of handling systems — flues and

8/1 Improvement of an existing brick chimney (viewed from above) by erecting within it individual insulated steel liners to serve two large steam boilers. Uninsulated oversized chimneys, with high heat losses, often give rise to acid condensation, with resulting corrosion and smut emission, especially when the user operates the plant at high efficiency. Good chimney design and installation are an important part of energy conservation.
Energy Equipment Co. Ltd, Leighton Buzzard, Bedfordshire.

ducts, as well as chimneys — internal surface temperatures must remain above the dewpoint of the gases (*see* Section 4.3). If the fuel burned contains sulphur, the *minimum* temperature attained, even at low load, must be above the acid dewpoint if acid deposition and corrosion are not to occur.

Improving the efficiency of all forms of fired heating plant introduces a risk of incorrect flue gas disposal. The desire to extract the last joule should be tempered with caution. Flue gases should be kept hot and cold air ingress, particularly through draught stabilisers, should be minimised, if loss of buoyancy and condensation are to be avoided. Modern dampers fitted to steam boilers can reduce heat losses considerably during shutdown or intermittent operation and the effectiveness of existing dampers should be examined. Ducts and flues often have large surface areas and numerous joints subject to thermal expansion and contraction. Good sealing and insulation are therefore essential. Redundant ducts and flues should be removed or effectively sealed. Fortunately, gas handling and chimney design are now well understood and reputable chimney manufacturers and erectors are well aware of the requirements for satisfactory performance.

The cost of moving large quantities of flue gases or other waste gases can be reduced by:

Making good use of chimney buoyancy.

Eliminating unnecessary pressure restrictions (e.g. old dampers and sharp bends).

Using fans that closely match the maximum requirements of the system and varying their output to match demand.

Keeping fan blades clean, and regularly cleaning filters and cyclones.

- **Keep flue gases above their dewpoint or acid dewpoint temperatures.**
- **Prevent cold air ingress to flues and fit modern dampers at critical points to reduce heat and pressure losses.**
- **Reduce pressure restrictions in all flue gas handling systems.**
- **Match fans to the duty required and keep fan blades clean.**
- **Regularly clean filters and cyclones.**

8.2 Heat recovery from waste gases

The *direct* use of exhausts and waste gases should always be considered before other heat recovery methods are examined, such as partial recirculation in drying applications (*see* Section 7.7), or lower-temperature drying or preheating. When hot, or even warm, waste gases are clean and still have a high

8/2 Modern boiler isolators with excellent gas tightness, no leakage around pivots and automatic control using geared electric motors have a distinct energy-saving value compared with simple dampers. With fluctuating loads, constant draught conditions can more readily be achieved without the admission of cold air, which can lead to corrosion and smut problems. During periods when the boiler is not being fired — for minutes, overnight or for a weekend — heat losses and heating-up times on restarting are cut considerably.
Heatlock Dampers, Croxley Green, Hertfordshire.

---

oxygen content (e.g. gas turbine exhausts), their use as preheated combustion air should seriously be considered.

Three important criteria should be met before *any* form of heat recovery is undertaken. First, the necessity for the surplus heat should be challenged. There is little point in installing an expensive recovery system if the surplus can be reduced by more efficient operation in the first place. Secondly, there should always be an appropriate matching use for the recovered heat. Thirdly, recovery should be economically viable and the recovered heat should not displace a cheaper source.

In many furnaces, in which the flue gases must necessarily be discharged at high temperature, heat recovery is well-established practice and is undertaken for two reasons:

To improve combustion and to obtain high flame temperatures by preheating the combustion air, especially when low-grade fuels such as blast furnace gas are used.

To improve efficiency by reducing the amount of heat finally discharged to the atmosphere.

Various methods of recovering process heat are available. Recovery may be by means of load recuperation (stock preheating) as in shaft furnaces (e.g. lime kilns and blast furnaces), by stock cooling or partial recirculation as in some ceramic kilns and driers, by producing steam or hot water in waste heat boilers or economisers, or by preheating combustion air using regenerators (intermittent heat transfer from intermediate heat storage) or recuperators (continuous heat transfer).

In new installations the precise type and size of heat recovery equipment and the materials of construction require most careful consideration, account being taken of waste gas volume, temperature and composition, usable recovered heat and economic life required.

The atmospheric conditions, operating temperatures and materials of construction of heat recovery equipment need clear definition. For example, alternating oxidising and reducing conditions (which may occur with slow response to controls) should be avoided in metallic recuperators, otherwise high rates of corrosion may result. To avoid condensation it may be necessary, at start-up or during low-load operation, to bypass heat recovery equipment until waste gases have risen to well above their acid dewpoint temperatures. Positive steps must be taken to prevent local or general overheating of heat recovery plant.

Existing heat recovery equipment needs adequate supervision and, like all energy plant, it should be properly instrumented if benefits are to be maintained. In particular, all ducts and flues should be gas-tight, heat exchange surfaces clean, and all fans clean and balanced. Significant changes in flow rates or temperatures should be investigated.

- **Reuse hot or warm waste gases directly, whenever possible.**
- **If waste gases cannot be used directly, recover heat through a heat exchanger but first check that:**
  the surplus does not arise from inefficiency
  there is a use for recovered heat
  recovery is likely to be economically sound.

8.3 Recovery methods

Regenerators and recuperators used for large, high-temperature furnaces are an essential part of their basic design. The requirements are well understood and documented. The brief descriptions below are of some specific types of heat recovery equipment of wider application, particularly equipment that may be added to existing plant. There are, of course, very many well-established types of industrial heat exchanger that may be used for heat recovery.

Waste-heat boilers

Heat recovery using waste-heat boilers is often applied to furnaces but their use is also common for engines (especially marine engines) and other plant whenever gas temperatures exceed about 260°C (500°F). In waste-heat boilers, heat transfer is mainly by convection and the design is therefore quite different from that of a conventional fuel-fired boiler in which a high proportion of heat transfer is by radiation.

The surplus heat available must be accurately assessed from the weight, temperature and specific heat of the gases being discharged. Account must be taken, too, of any physically active (e.g. abrasive dust) or chemically active (e.g. sulphur oxides) constituents. There must, of course, be a corresponding demand for heat or energy nearby at a time when the surplus heat is available. A clear idea must also be formed of the temperature at which recovery is to take place. This will depend on the proposed application. For example:

low-pressure hot water	82°C (180°F)
process steam at 700 kN/m^2 g (100 lbf/in^2 g)	170°C (338°F)
steam for power generation at 6300 kN/m^2 g (900 lbf/in^2 g)	280°C (534°F)

A greater proportion of heat will be recoverable at lower temperatures and high-temperature recovery will be more costly. This information is essential for the specification of a suitable waste-heat boiler and for a proper economic appraisal to be undertaken.

Self-recuperative burners

Heat from waste gases may be recovered directly in the firing process using self-recuperative burners in which combustion air may be preheated to about 650°C (1200°F). A high-velocity burner, recuperator and flue are combined in a self-contained, compact unit. The burner is particularly suitable for application to smaller batch furnaces — forge furnaces, heat treatment furnaces, intermittent kilns, etc. — where the potential for heat recovery is greatest. Many small forge furnaces lose 70% or more of their heat input in the flue gases.

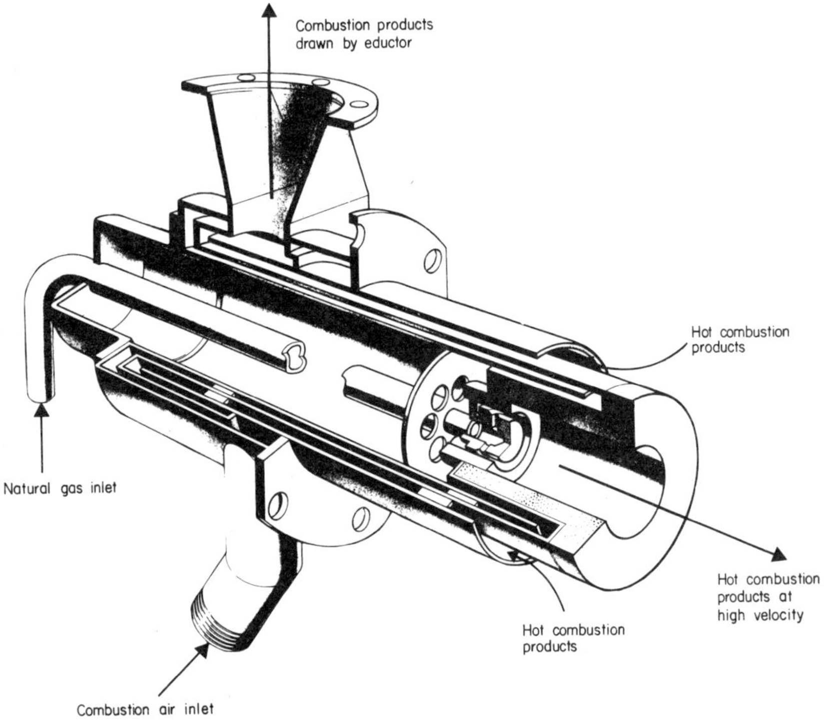

8/3 A recuperative burner, consisting of a 'tunnel'-type gas burner surrounded by a counterflow heat exchanger for preheating combustion air with outgoing hot combustion products. Recuperative burners have been applied successfully to various furnaces and kilns (especially types without regenerators or recuperators) to effect fuel savings and improve performance.
British Gas, London W1AA 2AZ.

Rotary regenerators (thermal wheels)

Rotary regenerators use an old-established principle developed by Ljungstrom and first applied to large power plant. Modern versions are available in a range of sizes and are suitable for recovery from process heating and air conditioning systems. A drum containing appropriate heat transfer surfaces is slowly rotated between two separately ducted streams, absorbing heat as it passes through the exhaust and transferring it to the incoming air as it passes through the inlet section. Efficiency is highest when the exhaust and inlet mass flows are similar.

8/4 A single-wheel rotary regenerator for recovering heat from industrial waste gases or ventilation exhausts. Exhaust and inlet streams are ducted counterflow through separate halves of the unit, heat being picked up by the rotor as it passes through the exhaust half and transferred to the cooler fresh air in the inlet half. The size and running costs of air-conditioning plant can also be reduced by using regenerators to precool incoming air using outgoing cooled air.

A range of sizes and rotor materials is available to suit various applications.

Curwen & Newbery Ltd, Westbury, Wiltshire BA13 3DZ

8/5 Heat recovery from continuous washing effluent, using a 90 litre/min heat exchange unit. This preheats washing water from 4°C to 77°C, reducing steam requirements and reducing the effluent temperature from 93°C to 21°C.

The APV Company Ltd, Crawley, West Sussex.

Cross-contamination can be controlled to low levels, if necessary, by suitable purging, though this lowers efficiency slightly.

Plate heat exchangers

Plate heat exchangers are widely used for heat recovery in many industries, especially for liquid to liquid systems. Batteries of various sizes can be built from modules. The plates, of materials to suit the fluids being handled, can be formed to produce large surface areas and high turbulence.

8/6 A stainless steel recuperator in which water vapour in the flue gases is condensed by direct water washing to recover both sensible and latent heat. Particularly suitable for gas firing. *Spur Engineering Ltd, London SW19 3HT.*

Run-around coils

Run-around coils are mainly applied where the surplus heat is not adjacent to the point at which it can be used. Heat exchangers, usually finned coils, are fitted in the exhaust duct and the supply duct and are connected by a pumped water circuit.

Heat pipes (thermosiphons)

The heat pipe is an elegant method of effecting heat exchange, heat being absorbed at one end of a sealed pipe to vaporise a fluid which, on condensing at the other end, gives up its heat. The fluid then returns to the vaporising end by capillary action through a woven metal 'wick' to complete the cycle. There are no mechanical moving parts. Normally heat pipes are used in the near-horizontal position and if required, heat transfer can be closely regulated by altering their angle of inclination. As with rotary regenerators, exhaust and inlet streams must be adjacent so that the heat pipe can span them both. Heat pipes are suitable for heating, cooling and for equalising temperature differences.

Heat pumps

The heat pump is of particular significance where the heat available for recovery is at too low a temperature for it to be useful if recovered by normal heat exchange methods. It can be used to extract heat from a low-temperature source and release it to a higher-temperature sink, using a vapour compression cycle. Work must be done to achieve this, but the heat pump produces an output about three times this input, which is often from an electric motor, though other 'prime movers' can be used. Because industry often has available *regular* low-grade heat sources (e.g. cooling water and effluents) which can be upgraded using heat pumps, industrial applications may have better economic potential than domestic ones.

The most promising applications are those in which heating and cooling can be used in combination and where the temperature differences involved are small. As well as straightforward heat recovery the heat pump can usefully be applied to drying applications.

- **Consider the whole range of recovery methods available before selecting the most suitable.**
- **Pay particular attention to waste gas or liquid temperature, volume and composition, and to likely operating conditions.**
- **Consider how the surplus heat and the application for recovered heat can be satisfactorily matched and brought closer together.**

8.4 Non-fuel (renewable) energy sources

Considerable development effort is being applied to devise practical systems to harness alternative, renewable sources of energy. Some of these are already finding practical application, though often on a small scale or in situations where financial viability is not of prime importance.

Wind power, geothermal energy and wave power are unlikely to be used on a large scale by industry and commerce in the near future, except perhaps as part of a local development project. Solar energy applications appear to have greater potential, especially for heating purposes. Equipment is generally simple and well developed. In industry and commerce, applications, which must usually be economically viable, are more likely to be successful where there is a requirement that matches solar heat availability, so that storage requirements are reduced (e.g. holiday camp pool and kitchen water heating), or where the end use is relatively insensitive or can accept heat intermittently (e.g. boiler feedwater preheating).

For general space heating and hot water applications, the greatest potential lies in *new* buildings where suitable aspects can be chosen and solar collectors can be incorporated into roof designs without great additional expense.

- **Take particular note of any summer process heating requirements and loads that are not time-sensitive and consider the possible application of solar heating.**
- **In new buildings, examine the possible incorporation of solar collectors.**

9 ELECTRICITY

9.1 Costs and tariffs

Electricity is generally regarded as an expensive form of energy. The overall efficiency of public generation from fossil fuels is usually low — 30% or so — so that electricity may be regarded, also, as wasteful of energy resources unless generation is combined with use of the heat that is otherwise wasted in a district heating or combined heat and power scheme.

However, the fuels used by power-stations are often low-grade ones and advantages in electricity end-uses — ease of control, high thermal efficiency and freedom from pollution — should also be taken into account in arriving at a balanced judgement in any particular case. When energy costs are a small proportion of total costs, reduced labour requirements, improved product quality and better yields may be significant and when any of these can be attributed to the use of electricity, its higher cost may well be justified.

For heating purposes, electricity is very versatile and in addition to indirect resistance heating (using separate elements and heaters), heat may be transferred by direct resistance (passing a current *through* the product), electromagnetic induction, arc or plasma-arc, dielectric loss and microwave absorption.

Whatever uses electricity is put to, controlling and reducing its cost begin with detailed knowledge of the tariff structure, adequate metering to all main areas of use and attention to three main aspects of utilisation:

Reducing the total amount of electricity used by greater efficiency and the eradication of unnecessary uses.

Developing a more uniform load, in order to reduce the maximum demand and improve the load factor.

Examining the application and performance of electric motors and other 'reactive' loads in order to improve power factor.

Maximum demand is the maximum hourly rate of consumption, based on a stated recording period, usually half an hour, within any accounting period. It represents the undesirable peak for which the electricity supplier must provide extra generation and distribution facilities — and charge for them accordingly. It is usually measured as kilovolt amperes (kVA). Load factor is the ratio of

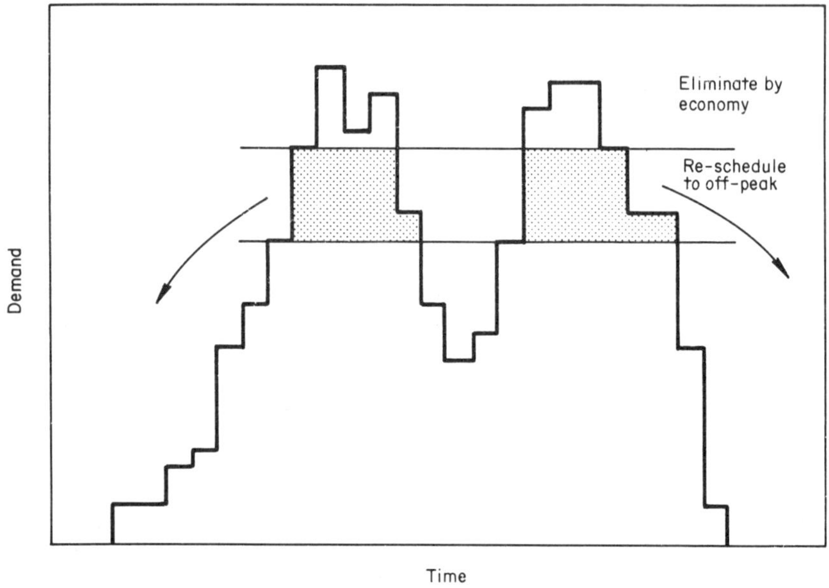

9/1 Reducing maximum electrical demand and improving load factor.

units actually used to the units which would have been used had the maximum demand rate been maintained continuously throughout the accounting period.

Part of the current required for alternating-current machinery is used to provide the magnetic field for the motor and it produces no direct mechanical work. Similar 'wattless', 'idle' or 'reactive' currents occur in transformers and chokes. The product of the total current, measured by an ammeter, and the supply voltage gives the apparent power used (in kVA). This is greater than the true or useful power (in kW) which is indicated by a wattmeter.

$$\text{Power factor} = \frac{\text{true power (kW)}}{\text{apparent power (kVA)}}$$

Supply costs depend to some extent on the capacity required to provide and distribute the 'idle' current and surcharges related to power factor may therefore be applied. When maximum demand is measured in kVA, an automatic penalty is imposed for power factor.

Because charges may depend on the maximum demand, load factor and power factor, as well as on overall consumption, each of which is considered below, it is vital to ensure that the most appropriate tariff is being applied from the point of view of the user, who should then plan his electricity use in a way that will further minimise his costs. Regular analysis of the various cost elements indicated in accounts is essential.

- **In considering the performance of all electrical plant and possible changes, seek the advice of a qualified electrical engineer.**
- **When considering a choice of energy source assess the possible benefits of using electricity as well as the prime energy cost.**
- **Select the most appropriate electrical heating method.**
- **Meter electricity to all main areas of use and keep records of consumption.**
- **Ensure that you are using the tariff best suited to your needs.**
- **Study the electricity tariff carefully. Plan to avoid surcharges and penalties.**

9.2 Reducing overall consumption

The effective capacity of any distribution system is improved by reducing consumption and ways of doing this should always be sought before extending an apparently overloaded system. To minimise distribution losses, transformers should be located as close as practicable to heavy loads.

As well as switching off items when not actually in use, using time switches, spring-loaded and foot-controlled switches and similar devices, every effort should be made to get more useful work — illumination, heating, cooling, etc. — out of every unit of electricity consumed.

Process heating

The most appropriate heating method should be selected and a check made to ensure that, for the particular application, electrical heating has a distinct advantage over fuel firing, especially if it contributes to peaks in demand. As with fuel-fired applications, proper attention is required to thermal insulation, thermostatic control and to optimum process times and temperatures. Proper planning to minimise preheating times and to ensure full loads is essential for economical operation.

Motive power

Unnecessary materials and personnel movement by lift, hoist, conveyor, escalator, etc., should be eliminated, full use being made of gravity whenever possible. Attention to the final operations for which motive power is required may also yield useful savings. Sharp tools, correct lubrication and aligned bearings in good condition can reduce frictional losses significantly.

All electrically driven plant should be loaded to its full capacity. Compressors are particularly inefficient at low load and smaller ones should be used for particular intermittent work, the main compressor plant being reserved for steady, full-load operation. Idle running, with no useful work being performed, still requires power to overcome friction.

The lowest speeds which will give satisfactory performance on plant such as mixers, rotary driers, pumps and conveyors should be used, together with the most efficient form of direct drive and correctly matched motors. Not only will power be saved but noise levels and maintenance will also be reduced. Power requirements are roughly proportional to speed for these types of application, but for fans and blowers the power requirements are roughly proportional to the cube of the speed for free discharge and the square of the speed for discharge throttled by dampers or valves. Operation against high back-pressures is therefore undesirable. Variable-pitch fans or speed control should be used to vary air flow rates whenever practicable. Cheap fans and blowers are often inefficient and the use of equipment of good aerodynamic design with known characteristics which can be properly matched to the duty required will effect long-term savings.

Illumination

There is ample evidence that good lighting improves work performance and reduces errors. Illumination of workplaces should therefore be to proper standards and preferably in accordance with some recognised code of practice (e.g. in the UK, section B9 of the CIBS Code). The important thing is that the level of illumination should be *appropriate* to the activity being carried out.

In storage, general recreational and similar areas, lower levels of illumination may be acceptable, provided work performance, security or safety standards are not sacrificed. Non-essential or decorative lighting and signs, especially in non-working areas, should be removed. In fluorescent multi-tube fittings, the whole unit should be disconnected — individual tubes should not be removed. Expert advice should be sought.

Everyone should be encouraged to switch off lights when they are not required. Particular care is needed, especially in office blocks, to switch off lights overnight. Office cleaning should be phased so that the whole building is not a blaze of lights, and consideration should be given to cleaning in daylight, or at weekends.

More switches, enabling smaller groups of lights to be controlled, may assist. In many instances, lights are left on unnecessarily simply because a few, controlled by the same switch, are needed in one dim corner. Time controls and photocell control (with manual override) may be useful in some cases where zones of fairly uniform illumination can be defined and the number of lights justifies the cost. Photocells should be shielded from direct sunlight and exterior ones should be heated or protected to prevent frost and snow causing malfunctioning.

In new buildings, lighting can be considered as part of a properly integrated design, together with heating, sound and other factors. In existing buildings this is not usually possible and where high levels of illumination are necessary, the contribution to space heating should not be underestimated. Other conventional heating should be reduced to compensate in preference to overventilation.

Windows and roof lights are sources of high heat loss from buildings so that if full advantage is not taken of the natural light which they provide, they may as well not be there. They should be kept clean and unobstructed and surrounds and sills should be painted in light colours. Light-coloured decoration improves all lighting, natural and artificial. Planned cleaning of lighting fittings should be carried out and poorly designed or ineffective shades, diffusers and reflectors should be removed or replaced. The light output of all lamps decreases with time and, in suitable circumstances, consideration should be given to planned mass replacement towards the end of their rated life, rather than the renewal of individual lamps when they fail completely.

Although some savings may be effected by reducing unnecessary illumination, the main savings can usually be made by replacing tungsten filament incandescent lamps by high-efficacy fluorescent tubes or high-pressure discharge lamps. Up to two-thirds of the annual cost of lighting may be for electricity alone so that the cost of changing over can often be quickly recovered. High-pressure sodium (SON, SONT) lamps have especially high efficacy but they are unsuitable where true colour rendering is required. In such cases mercury halide (MBI) lamps may be more suitable. A replacement programme should be undertaken only after taking expert advice, especially if rotating machinery is involved, or faithful colour rendering is important.

Exterior lighting may be required to enable useful uncovered areas to be used for work, storage, parking or recreational purposes. Exterior lighting may also be needed for security reasons. Each use merits individual consideration but a reduction in overall levels of illumination should be sought in combination with rescheduling or shortening of periods of use. Where floodlighting, illuminated advertising signs and exterior decorative lighting are used, their precise purpose should be defined, together with their periods of use. Their impact may well be greater with more selective use (e.g. on fine evenings only and not in bad weather).

- **Switch off electrical plant and lighting when not in use. Make this easy by having good switching arrangements and use time switches and other automatic methods.**
- **Examine applications of electric power and ensure that there is efficiency at the final point of use.**

- **Ensure that plant is fully loaded. For moving plant, use the lowest speeds that will do the job satisfactorily.**
- **Examine the characteristic curves of fans and blowers, and select them carefully. Control fan outputs by varying blade pitch or speed of rotation and avoid the excessive use of dampers and valves.**
- **Reduce illumination in storage and non-working areas and review all exterior lighting.**
- **Reduce space heating input to compensate for heat from lighting and other equipment. Do not overventilate.**
- **Use existing natural light sources to the full. Keep windows clean and unobstructed. Use light-coloured decorations.**
- **Clean light fittings regularly and check the effectiveness of diffusers, reflectors, etc. Consider planned replacement of lamps and tubes.**
- **Replace tungsten filament lamps by tubes and lamps of higher efficacy. Seek expert advice.**

9.3 Load factor improvement

After all unnecessary uses of electricity have been eliminated, the next step is to reduce the peaks and generally spread demand. Adequate metering and records are essential to establish the load profile and to identify the peaks. Systematic listing of all items of plant and equipment that use electricity will assist the

---

9/2 The electronic display of a maximum-demand load controller that indicates elapsed time, present demand and predicted demand compared with a predetermined target.

The unit monitors consumption against manually or automatically set targets to provide early warning of load shedding well before the target is exceeded.

Horstmann Engineering Products Ltd, Bath, Avon.

identification of offending items. In spreading loads, advantage should be taken also of night tariffs, rebates and other special financial incentives.

General economy in use is especially important at peak periods, non-essential equipment being switched off as soon as a predetermined load level is approached. The setting of a demand target which must *not* be exceeded is essential. Alarms and manual switching-off of preselected items may be used, but more sophisticated controls are now widely available that predict peaks and automatically shed loads.

The main improvement will probably come from the planned use of major items of electrical plant. The heating-up of electric furnaces should be staggered and the starting (or restarting) of more than a certain number of intermittently used large motors in any demand period should be avoided. It should be noted that meter reset times for maximum demand do not necessarily fall tidily on the hour or the half-hour, and they may be changed from one accounting period to another.

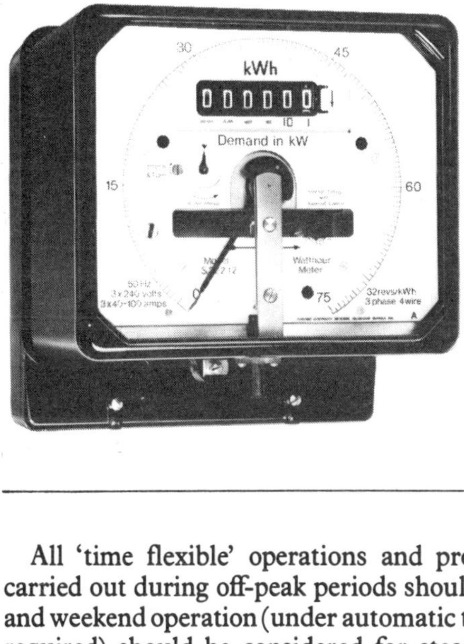

9/3 A maximum-demand meter. As well as indicating total consumption in kWh, a demand pointer is advanced by a pusher arm to indicate the average kW demand during the predetermined time interval, usually 30 minutes. The pusher arm automatically resets to zero at the end of each period, leaving the demand pointer at its most advanced position. *Sangamo Electricity Metering, Felixstowe, Suffolk.*

All 'time flexible' operations and processes which can conveniently be carried out during off-peak periods should be rescheduled accordingly. Night and weekend operation (under automatic time control if close supervision is not required) should be considered for steady continuous operations requiring motive power like grinding, polishing, pumping and mixing and for more straightforward process heating like annealing, enamelling, drying and pre-heating. Battery charging and storage-heater regeneration should be carried out at night. Possible side-effects should be considered. There is little point in rescheduling an operation to a night shift if a host of other services have to be provided in order to do so.

Where processes, especially drying, can only be carried out at peak times, non-electrical methods should be considered. Whenever electricity is used for water heating or chilling, sufficient storage should be arranged to meet peak demands. Tanks should, of course, be properly insulated.

The on-site generation of electricity is considered briefly in Section 10.1. Many organisations have existing standby generators for emergencies and although these may not be suitable or, without heat recovery, economical if used continuously, their use to meet the worst peak in demand or for specific heavy intermittent loads should at least be considered. Generators 'turned over' periodically in this way are more likely to be 100% reliable when required in an emergency.

- **List all items of plant that use electricity. Identify those that contribute most to peak demand.**
- **Encourage economy at peak periods. Set a demand target which must not be exceeded and switch off selected plant as the target is approached.**
- **Stagger the start-up times of large motors and the heating-up times of furnaces. Check the maximum demand meter reset time to determine the optimum programme.**
- **Reschedule suitable processes to off-peak periods where possible. Check that no additional costs will result.**
- **Consider non-electrical alternatives where processes can only be carried out at times of peak demand.**
- **When electricity is used to heat or chill water, do this at off-peak periods.**
- **Use existing standby generators to meet peak demands, or part of the electrical load on a regular basis.**

9.4 Power factor improvement

The power factor of an induction motor is highest at about full load. Overall power factors can therefore be improved by keeping all motors as near fully loaded as possible. Care should be taken, however, not to *over*load motors as this will result in loss of efficiency and overheating. A large underloaded motor may still be more efficient than an overloaded small one.

The current surge on first starting a motor must be minimised and the current required to bring a motor (especially one driving heavy rotating plant) up to its correct operating speed may be several times that required to maintain it once acceleration is completed. The initial selection of a motor size, and also its connections and starting gear, are therefore very important. Various codes and standards such as BS 5000 and IEC 158 give useful guidance. As individual power factors must inevitably be low during acceleration, the staggering of the starts of large motors is desirable.

The simultaneous pressing of dozens of starting buttons at the beginning of the working day may be something that can be avoided, but even under steady running conditions, power factors may still be unacceptably low and other processes and uses (e.g. arc welding) may contribute. The use of alternative types of motor or the use of capacitors in parallel with the loads can effect improvements.

Capacitors require negligible maintenance and have no running costs. However, it is uneconomic to try to correct a power factor to unity and 0.95 is usually an acceptable target. The cost of installing capacitors and the time which will be taken to recoup this can only be considered after a survey of power factor. Reputable capacitor manufacturers will undertake such surveys and advise on installations. For any change of this sort a qualified electrical engineer must always be consulted. Power factor correction has the incidental advantage of improving the useful capacity of the existing distribution system at no additional cost. This may be helpful where a system has reached its limit.

- **Select electric motors so that they run as near fully loaded as possible.**
- **Select starting gear for motors to give optimum performance and to maximise power factor as far as possible.**
- **Stagger the starting of electric motors.**
- **Investigate the possible use of alternative types of electric motor. Seek expert advice.**
- **Improve power factors by installing suitable capacitors. Seek expert advice.**

10 SERVICES

10.1 On-site electricity generation

For many users, the public electricity supply is a very satisfactory and reliable source of power. But on-site generation should be considered if it is essential to have power on the rare occasions when the public supply is cut off, or if there is a suitable balance of power and heat requirements to justify heat recovery from the engine.

The purpose of on-site generation should be established quite clearly at the outset. If the generator is required only to meet essential electrical demand in an emergency, the first consideration is to ensure that it is maintained in perfect running order so that it will start immediately and run reliably when needed. It

10/1 A small 21 kVA generating set which supplies all the emergency power requirements of a small engineering works. Most large boilers, furnaces, kilns, etc., require electricity for pumps, fans and controls and provision of minimum standby supplies should be considered for such plant to prevent shut-down in an emergency.
Dale Electric of Great Britain Ltd Filey, North Yorkshire.

is unlikely that anything more than the most rudimentary heat recovery will be justified on such standby plant. If, on the other hand, it is desired to cut costs by generating on a more continuous basis, or regularly to 'shave' electricity demand peaks, then heat recovery is probably essential. In this case the engine should be selected to give the electrical output required but *also* the most suitable balance of power and heat; that is, to provide the basis for a total energy system.

'Total energy' generally means the on-site production of electrical or mechanical power, with maximum heat recovery. Heat from on-site generation may be recovered in various ways:

Passing engine exhausts through suitable waste-heat boilers to produce hot water or steam.

Utilising heat from engine and oil cooling systems.

Direct heating or drying where exhaust gases are suitable, or heating air indirectly where they are not.

Using exhausts from gas turbines as preheated combustion air where they have a sufficiently high oxygen content.

Using steam from steam turbines for process heating at lower pressure.

Engines

Generation with steam turbines is widely practised throughout industry. In the UK, 20% of industrial power is produced in this way. Back-pressure turbines are preferred by many users, the reduced-pressure steam which passes from them being used for process heating. As the efficiency of actual generation is not very high compared with public generation, the heat remaining can hardly be described as 'waste heat'. Steam heating requirements may, in fact, dictate the amount of generation possible. The overall system design, and selection of the most appropriate types of turbine, is a matter for the expert.

Diesel engines, dual-fuel engines (gas engines with oil injection to assist ignition) and spark-ignition engines all have relatively high power generation efficiencies (approximately 32–40%) and heat may also be recovered from the cooling systems and exhausts. Capital costs are low and reciprocating internal-combustion engines are well-established and reliable, though maintenance costs are relatively high. Heavy oil may be used for larger, slower-speed, diesel engines.

Although they require higher-quality fuels and are less efficient power producers (approximately 17–28%) than reciprocating engines, industrial gas turbines have some compensating advantages. Clean exhaust gases at high temperature (500°C) may be used for direct drying applications or for producing steam or hot water. Because they usually have a high residual

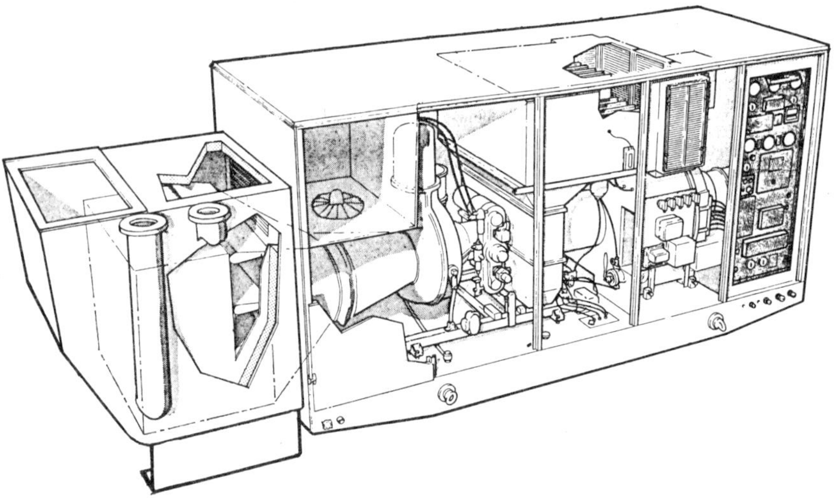

10/2 Using the exhaust of an industrial gas turbine generator to heat water can raise the overall efficiency (on gross CV) to over 75%. The total energy unit illustrated is a purpose-built package, complete with controls. The generator will provide 550 kW of electric power and the waste heat boiler an output of 2200 kW (7.5×10^6 Btu/h). *Centrax Ltd, Newton Abbot, Devon.*

oxygen content, gas turbine exhausts may also be used as preheated combustion air in suitable applications, and with varying turbine load, direct auxiliary firing may be introduced to maintain or boost the exhaust gas temperature.

Multiple-unit installations to enable varying demand to be met and to provide capacity for maintenance and standby are common for all internal-combustion engines. Back-pressure steam turbine systems more often consist of large turbines without standby capacity, arrangements being made for any emergency power requirement to be drawn from the public supply.

Diesel engines are usually selected when maximum electricity generation is required and the opportunities to use recovered waste heat are limited. Where a high proportion of energy input can be utilised as clean high-temperature gases, gas turbines are a more obvious choice. Where there is a large demand for process steam (or waste steam is available very cheaply) a steam turbine system is more likely to meet the requirements.

In selecting the most suitable prime mover for any particular situation, the precise power and heat requirements should be carefully assessed and compared with engine performance figures over the range of operating conditions likely to occur in practice. As well as capital, fuel and lubricant costs, the additional cost of skilled labour for maintenance, spares, additional management, public supply back-up, fuel stocks, etc. should be fully considered. The desirability or otherwise of a long-term commitment to a particular piece of generating plant and possible reduction in future flexibility must also be taken into account. Provision must be made for regular planned maintenance and overhauls.

Standby generation and peak lopping

If standby plant is used solely to provide power requirements in an emergency, economic evaluation is difficult. Costs should be regarded as insurance and weighed against loss of production, waste of product, loss of customer confidence, work-force lay-offs and so on. If sensitive items of equipment (e.g. computers) are involved, particular attention should be paid to the time delay

10/3 Two 394 kVA generating sets which automatically provide peak lopping facilities as well as mains failure standby for a continuous process plant. For peak lopping, either one or both of the sets are started, depending on the load demand, the sets being automatically synchronised and connected in parallel with the mains supply. *Dale Electric of Great Britain Ltd, Filey, North Yorkshire.*

which can be tolerated between loss of the public supply and connection of the standby supply.

An industrial generating system without heat recovery cannot normally compete with a public supply as far as average cost per unit is concerned. But industrial electricity loads are not constant and where a tariff includes heavy maximum demand penalties, partial on-site generation to meet the expensive peak demand may reduce the total cost of power even without heat recovery. Generation plant intended primarily for emergency standby use may be utilised, with the provision of additional control gear and means of safeguarding the public supply.

- **Establish clearly whether the purpose of on-site generation is to provide electricity in an emergency or to cut costs.**
- **For anything but emergency generation, consider what use can be made of waste heat and select a prime mover that will provide a suitable generation/heat balance and the right sort of heat.**
- **Where electricity generation requirements are high, consider reciprocating engines.**
- **Where electricity generation requirements are not high but use can be made of hot clean gases of high oxygen content, consider gas turbines.**
- **Where there is considerable demand for low-pressure steam, consider back-pressure steam turbines, especially if high-pressure steam is available. If boilers are due for renewal, consider using higher pressures and generating.**
- **Include in the evaluation the other additional costs not normally incurred when using the public supply. Allow for variations in load, breakdowns and regular maintenance.**
- **Regard emergency generators as insurance and evaluate the cost and probability of stoppages and shut-downs caused by power failures.**
- **Consider adapting standby generators for regular peak lopping.**

10.2 Compressed air

Air is a plentiful commodity and, at atmospheric pressure, it is free. Compressing it is an expensive business, however, and the efficient production and use of compressed air can yield valuable energy savings. An attempt should be made by every user to assess, however approximately, the cost of his compressed air.

Compressors

Air should not be compressed more than is necessary. About 15% more power is required to compress air to 7 bar g (105 lb/in^2 g) than to 5 bar g (75 lb/in^2 g). An optimum pressure for each system should be selected with care, the lowest pressure compatible with correct operation of the equipment served being determined with allowances for losses in the distribution system. The use of lower pressures will also reduce leakage.

Compressor selection should be based on the pressure required and the average and maximum loads (including future loads) to be met. Two-stage or multi-stage compression is generally more efficient, especially where air is cooled between stages (intercooling). If the volume requirement for low-pressure air is large (e.g. for extensive instrument control), consideration should be given to providing a separate low-pressure system. Where particular items of equipment require higher pressures than the rest, further compression of the main air supply near the points of use or separate compression should be considered.

Main compressors should operate steadily at close to their rated maximum capacity but should not be overloaded as this may lead to higher maintenance costs. Where several compressors serve the same system, control should be such that as many as possible operate at nearly full load, with only one handling load variation. Oversized compressors operated at low load are very inefficient, and where any large items of equipment require air intermittently, even at the same pressure as everything else, consideration should be given to supplying them from separate compressors.

10/4 A compact, well-designed rotary air compressor, well suited for use with individual groups of machine tools, shotblast cabinets, etc.

Even a well-maintained compressed-air system wastes energy through friction, leakage and contamination. Decentralised air compression minimises these losses and further energy can be saved by shutting down individual compressors when they are not required, instead of operating a central plant inefficiently at part load.

The Hydrovane Compressor Co. Ltd, Redditch, Worcestershire.

Obviously, all compressors should be switched off when not in use and they should not be allowed to idle during meal breaks, shift changeover periods, etc. To assist the detection of deterioration in performance or abnormal running, simple records should be maintained of air pressure and temperature after each stage of compression, air delivered against power consumed and cooling water requirements. This should be done for full-load and off-load conditions.

Power requirements increase by about 1% for every 3°C increase in air inlet temperature. Cool, dry, clean air should be compressed and intakes should not be sited in hot, steamy or dirty positions. Intake filters should be regularly cleaned — well before dirt causes significant pressure restrictions.

When air is compressed its temperature rises and it will more readily hold moisture. This will condense out, however, in the distribution system as the air cools, causing corrosion and malfunctioning of equipment unless positive steps are taken to remove it immediately after compression. There are a number of ways of doing this, most of which use artificial cooling (aftercooling) to condense the water vapour. In critical applications, further chilling or chemical drying may be required but there is no virtue in cooling much below the minimum temperature likely to be reached in any part of the distribution system.

On large plant, heat gained by water used for compressor cooling should be utilised for example, for indirect boiler feedwater preheating. Heat removed in chilling should also be returned to the air, if possible, after the condensed moisture has been separated. Where mains water is used for cooling and is then run to waste, the amount should be strictly controlled with a thermostat, to limit unnecessary flow. Water should preferably be reused after cooling. Mechanical cooling or modern cooling towers are more satisfactory than cooling 'ponds'.

Air distribution

Air lines should be kept free from rust, moisture and excessive amounts of oil. Even with aftercooling, some water may still condense in the distribution system and it should be drained automatically from all air receivers and regularly from air lines. Drainage is facilitated if lines slope correctly towards drain valves, which should be kept firmly closed when not being used. Working points should themselves be positioned above the drain points. Lubrication points should also be provided as near to the working points as possible except for specific items of equipment (e.g. instruments) which require oil-free air. For all equipment, excessive lubrication should be avoided.

The *Guide to the Selection and Installation of Compressed Air Services*, published by the British Compressed Air Society, contains much useful information.

Leaks obviously waste energy, but they also reduce the effective capacity of compressor plant and may, in extreme cases, reduce significantly the perfor-

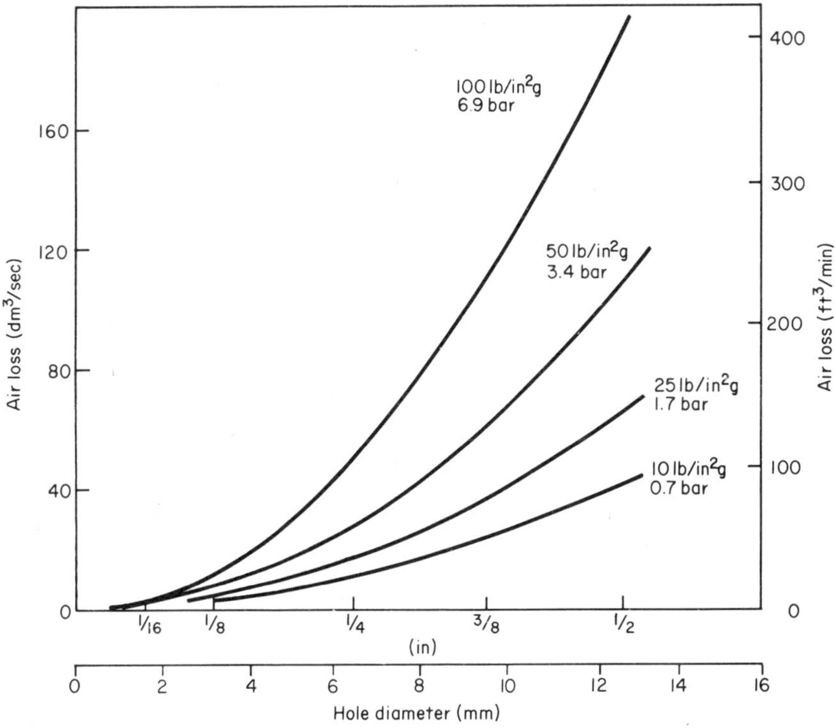

10/5 Discharge of air from compressed-air systems for various pressures. Air losses are measured at atmospheric pressure. Compressed air is one of the most expensive services to provide and minor leaks can quickly add up to a considerable loss.

mance of the equipment served. Leaks should always be rectified before any increase in compressor capacity is contemplated. A small amount of leakage is inevitable even in a well-maintained system, but any loss above about 5% should be regarded as unacceptable. Unfortunately, air leaks cannot be seen, smelt or, in many industrial situations, even heard so that they go unnoticed unless regular inspection is carried out, preferably during a shutdown period.

Unnecessary cooling of air lines should be avoided and when waste heat is readily available, reheating of air near the point of use will effect useful savings, though this should be undertaken with care. A 10°C increase in temperature will save about 3% of air.

Air-driven equipment

The use of compressed air as a continuous source of power should always be challenged and the economics of direct electric drive considered for such applications. For example, in hot working environments, after normal insulation and screening of hot surfaces have been attended to, electrically driven fans should be used for continuous cooling purposes. Compressed-air-operated exhausters, air movers and similar equipment should be reserved for specific local applications like fume dispersal, the cooling or ventilation of confined spaces, or the dilution of explosive atmospheres, for which they are well suited.

The quantity and minimum pressure necessary should be determined for all items of air-operated equipment. Manufacturers should be consulted and proper performance data obtained. Ideally, air should be available at the correct pressure required but, in practice, where a number of points on the same distribution system need different pressures, some form of pressure control or flow-limiting devices must be used, particularly for equipment like guns for swarf blowing. The operation of many pieces of paint-spraying equipment at lower pressures may well save paint and improve painting quality as well as saving air. The use of simple lever-operated valves in place of hand-wheel types encourages turning-off.

- **Determine the optimum pressure required for satisfactory operation of all items of equipment. Do not use higher pressures than necessary.**
- **Match compressors to pressure and volume requirements and avoid oversizing.**
- **Operate compressors at just below their maximum capacity, but do not overload them.**
- **Simplify distribution, maintenance and control by using matched individual compressors to serve particular groups of machines or tools.**
- **Provide separate compressors or boosters for items of equipment which operate intermittently or which require pressures markedly different from those needed by the main system.**
- **Switch compressors off when not in use.**
- **Use non-working periods to find and rectify leaks and to clean the system. Do not install further compressors without first thoroughly checking for leaks.**
- **Maintain simple records of compressor performance to enable faults to be detected quickly.**
- **Site air inlets in cool, dry, clean positions and clean filters regularly. Cool the air between and after compression stages.**
- **Recover heat from cooling operations and use some to reheat the compressed air, if possible.**

- **Keep air distribution lines clean, dry and warm. Avoid any accumulation of water or oil by sloping lines correctly towards accessible drain points.**
- **Challenge every use of compressed air, particularly for continuous applications.**

10.3 Water

Like air, water too is often taken very much for granted and little regard is paid to its cost or effective use. The precise quantity and quality of water required for every application should be examined to see if either or both can be reduced. With the increasing cost of water and attention being paid to effluent disposal, recycling should be practised whenever possible, especially if the water has been heated or treated in some way.

All uses of water for cooling should be reviewed to see if they really are essential and that the most effective method is being used. The speed of cooling should be questioned. Fast cooling may not be required if a process can be carried out earlier or if space is available for air cooling. The artificial cooling of any material which is later reheated should be challenged. Overcooling of engines and bearings of all sorts should be avoided and thermostatic and time controls can often usefully be employed to limit water flow. Rather than discarding cooling water, consideration should be given to reusing it. If this cannot be done, recovery of the heat should at least be examined before it is allowed to pass to the drains.

All treated water should be very carefully conserved and the amount of make-up minimised. Water is not a simple substance and whatever its source, a water to be used for industrial purposes is likely to contain impurities — dissolved minerals and gases as well as suspended matter. Before treatment, consideration must be given to the use to which the water will be put — cooling, washing, steam-raising or as a heat transfer or process material — and the treatment specified accordingly. The range of waters available and possible uses is wide and it is important that each use should be considered individually. Whilst requirements for particular alkalinity, hardness, solids levels, etc. should be assessed and met, care should be taken not to overtreat or carry out inappropriate treatment. Simple filtration or settling may be all that is required before reuse. Chemical dosing or other treatment should be carried out by a responsible person under expert guidance.

Inadequate treatment of boiler feedwater can have serious consequences as far as corrosion, scale formation (leading to reduced heat transfer and overheating) and steam purity are concerned. In extreme cases boiler failure may result. Modern boilers are much more compact and efficient than the Lancashire and Cornish boilers of a few years ago, and the feedwater requirements are also correspondingly much more demanding.

Water should be metered, not only at the supply point but also to each main area of use. Water meters are relatively cheap. As well as providing basic data to assist economy drives, metering also makes easier the regular checking for leaks in water distribution systems. Restrictors can be used to limit flow in known wasteful applications and unused sections of water main should be isolated and drained. The cost of water pumping can be reduced by operating pumps at close to their full rating and by keeping mains clean and unobstructed to minimise pressure losses. Water levels at pump suctions should be maintained as high as possible.

The surplus energy in water under pressure may sometimes be utilised in industrial situations. Large water turbines are usually associated with major power generation schemes but many mills still have small turbines in use and modern compact water motors are available down to very small outputs (fractions of kW and bhp). These are useful in flammable atmospheres or where variable speed is required.

- **Determine the exact quantity and quality of water required for every application. Ensure that treatment is properly carried out, but do not overtreat.**
- **Avoid unnecessary water cooling and check that the most effective cooling method is being used.**
- **Recycle water, especially treated water, whenever possible.**
- **Reduce water pumping costs. Create the maximum head of water at the pump suction and eliminate pressure restrictions.**
- **Meter water to all main areas of use and limit flow in wasteful applications.**
- **When surplus energy is available in water under pressure, consider using it to drive a water motor.**

10.4 Lubricants

Friction wastes energy. Attention to the correct lubrication of all moving parts not only saves energy but also increases the life of the equipment. Equipment manufacturers are usually very specific in their recommendations of lubricants and these recommendations should be adhered to. Without evidence that this may be done safely, the temptation to use lubricants for longer periods than those specified should be resisted. Such economies may prove false in the long run.

Economies should first be sought in the avoidance of waste and overstocking. Odd barrels and drums of 'special' lubricants are often wasted and an attempt should be made to rationalise stocks by standardising on a smaller number of high-quality multi-purpose grades which, at the same time, fully meet equipment manufacturers' recommendations.

Cleanliness is essential in the storage and handling of lubricants and they should be held in well-run stores. Advice on good practice is freely available from major suppliers. Barrels and drums should each be clearly identified and stored in properly constructed racks or on pallets to prevent damage and deterioration. Once opened, particular care must be taken to ensure cleanliness and correct labelling. Although the life of most lubricants is long, the packages themselves tend to deteriorate with exposure and rough handling and lubricants should be used in their delivered sequence, i.e. first in, first out. Positive steps should be taken to institute simple stores records and procedures, and to prevent pilfering, especially of grades suitable for private motor vehicles.

Some form of filtration or purification is an essential part of most lubrication systems. The filtration must be matched to the amount and nature of the contaminants. Both vary widely and should be assessed for each application before the most appropriate filtration system is selected. With the tendency in modern plant and equipment to operate at higher pressures and speeds and with closer tolerances, the demands made on lubricants and filtration are much greater than in the past.

Used lubricating and hydraulic oils are still valuable materials which can sometimes be reprocessed or reconditioned. This should be entrusted only to reputable and properly equipped organisations who will guarantee performance.

The use of so-called 'waste oil' as fuel is generally undesirable but if this is the only method of disposal it should be approached with caution. Waste oils may contain quantities of carbon and finely divided metals, including lead, and also be contaminated by motor spirit, diesel fuel or kerosine, which can introduce an explosion risk. They should be stored, and preferably used, quite separately from normal oil fuels and expert advice should be sought on their storage and application.

- **Lubricate in accordance with equipment manufacturers' or plant builders' recommendations.**
- **Avoid waste and overstocking by rationalising grades and having a well-run lubricants store. Keep simple records, and prevent pilfering.**
- **Avoid damage to packages caused by slovenly handling and storage. Use packages in sequence: first in, first out.**
- **Keep lubricants clean and correctly labelled, especially after packages have been opened.**
- **Ensure that each oil filtration system is entirely appropriate to the amount and nature of contamination.**

- **Consider the reuse of oils after reconditioning, but use only reputable reconditioners.**
- **If waste oil is used as a fuel, install a properly designed storage and handling system. Get expert advice. Never mix waste oil and ordinary oil fuels in storage.**

11 ROAD TRANSPORT

11.1 Driving techniques

Although the internal-combustion engine is highly developed, it is also inherently inefficient. When used for road transport this weakness is further aggravated because, in these applications, highly refined oil products are consumed.

Fuel savings may be made by paying attention to various aspects of vehicle design and maintenance, considered below. The greatest savings, however, are usually concerned with human rather than mechanical factors — in the actual driving of the vehicles and their operational management.

Before *any* possible savings are examined, a fundamental reappraisal is needed of what we mean by 'performance'. The Energy or Transport Manager should encourage all drivers, and office staff, to think of performance as *useful work per gallon*, rather than miles per gallon and certainly *not* as rapid acceleration or high top speed.

Saving fuel in transport depends to a great extent on the common sense and cooperation of every driver. Success is achieved only by frequent education and exhortation to emphasise the vital need for economy. Simple records of miles driven or loads carried compared with fuel used may assist control and the identification of poor drivers, but the *real* savings will be made on the road, away from close supervision and monitoring.

All drivers should be encouraged to observe some basic fuel-saving rules:

Avoid rapid acceleration and change into top gear as soon as possible, but do not let the engine labour.

Avoid harsh braking, which wastes fuel *and* damages tyres. It can be reduced by using moderate speeds, thinking and looking well ahead and not driving too closely behind the vehicle in front.

Keep tyres correctly inflated, but drivers of tractors and similar vehicles should check that wheel-slip, also wasteful, does not arise from *over*inflation.

Switch off the engine when waiting, loading or unloading. A battery in good condition and a well-tuned engine will encourage this.

Do not use the choke for longer than necessary. The engine should be warmed up by driving at reduced speed for a mile or two.

Cut down on weight by removing unnecessary tools and equipment and reduce drag by removing roof racks and ladder rails when not required.

If possible, manoeuvre the vehicle when the engine is hot.

Incentives can be used for drivers with good records of economy and safety. Safe driving is also economical driving and every encouragement should be given to responsible thoughtful drivers who practise safe driving techniques.

- **Ensure that everyone knows that performance means work per gallon, not miles per hour.**
- **Ensure that drivers know the main causes of high fuel consumption. Use posters, stickers and other reminders.**
- **Maintain simple records of fuel consumption related to mileage or preferably ton-miles.**
- **Emphasise the value of good driving techniques and encourage a pride in driving well.**
- **Encourage safe driving.**

11.2 Vehicle utilisation

The most effective way to save energy on the road is not to make the journey at all and a study of vehicle operations in any organisation should begin with a critical examination of the need for every journey. For many sales and service activities the telephone can be a satisfactory substitute. When the sale can be made or the problem resolved by phone, substantial savings will be made. Within factories, the use of conveyors, pipelines and pneumatic systems to move goods and materials may well reduce costs.

Many journeys *are* necessary, however, and savings then depend on how and when the journeys are made. To minimise mileage and at the same time avoid congested areas in towns and cities demands skill and knowledge, and the value of good route planning should not be underestimated. In organisations with large transport fleets, routeing can be assisted by computers and other sophisticated aids, but for many smaller operators there is no substitute for the experienced, intelligent planning supervisor. Planning can only be effective, however, if the vehicles available are properly matched to the loads which they are required to carry. They should not be overloaded but full payloads and back loads should be carried whenever possible.

The size and balance of any transport fleet needs frequent review. Replacement vehicles should be of lighter construction, if possible, and, for internal transport, the use of purpose-built fork-lift and other trucks will show benefits in fuel economy. Diesel engines should be used in preference to petrol engines for all but the smallest vehicles. Conversion of spark-ignition engines to liquefied petroleum gases has reduced overall operating costs through reduced maintenance and lower rates of duty, but, because of limitations on refuelling, the use of LPG is usually restricted to locally operating vehicles, such as delivery vans. The use of LPG also introduces a considerable weight penalty.

Warehouses and other storage areas should be sufficient to permit consignments to be consolidated. Delivering goods as soon as they are produced simply because there is nowhere to store them for a few days is unlikely to be very energy-efficient. Delivery priorities should be properly determined — urgent deliveries really should be urgent.

The suitability of company cars should also be reviewed, even if it is a rather delicate subject. The company's professional lorry drivers are unlikely to heed advice on saving fuel if senior management continue to be extravagant. Prestige cars should be reserved for prestige occasions and more intensive use should be made of multi-purpose vehicles like estate cars, preferably with regular drivers. Cars and lorries with air-conditioning systems and automatic gearboxes use more fuel than those without them. The extra initial and running costs of these refinements should be weighed against improved driver comfort and reduced wear on gearboxes, especially those of pool cars used by a succession of indifferent drivers.

- **Challenge the need for *every* journey.**
- **Consider alternatives to engine power for conveying goods internally.**
- **Put someone capable in charge of delivery planning and vehicle routeing. Ensure that sufficient planning aids are available.**
- **Frequently review the composition of the vehicle fleet so that vehicles are of the right type and size.**
- **Set up close liaison with the sales function to ensure that correct priorities are applied to deliveries.**
- **Consider greater stockholding of finished goods to smooth delivery patterns.**
- **Review the use of company cars and reduce the number and size by introducing multi-purpose vehicles as replacements.**

11.3 Vehicle design

In most countries there are legal requirements for the size, axle loading, power/weight ratio and other aspects of commercial vehicle design, so that opportunities for effecting greater fuel economy may be limited to some extent. Although there are serious and well-founded objections to very large vehicles passing through towns and villages with unsuitable roads, there is also ample evidence that they provide good fuel economy. When vehicles are purpose-built as for example oil tankers, there are often opportunities to reduce their unladen weight by the use of aluminium alloys, reinforced fibreglass and other lightweight materials for body, cab and engine construction.

Although basic designs of reciprocating engine appear to have limited scope for further development, the *selection* of the most appropriate power units and gearing can still yield useful savings. Again, the opportunities are greatest for purpose-built vehicles operating on long hauls, mostly in top gear. High-torque, low-revving engines have distinct advantages for motorway operation, provided the savings are not then dissipated on the return journey by an empty vehicle being driven at too high a speed.

11/1 Tractor units fitted with therostatically controlled radiator shutters. On the left, the shutter is closed to prevent the flow of cooling air. The engine on the right, having reached its operating temperature, has its shutter open.
Kysor Industrial (Great Britain) Ltd, Whitworth, Lancashire.

Keeping engines at their optimum operating temperatures can be achieved by the use of thermal clutch fans which operate only when a predetermined engine temperature is reached or by the use of thermostatically controlled radiator shutters or blinds. As with other devices, cost effectiveness is greatest when they are incorporated in new vehicles.

The reduction of aerodynamic drag, which increases as the cube of the vehicle speed, is of particular significance. At 100 km/h (60 mph), about half the power output of a lorry is needed to overcome wind resistance. Smooth-sided, round-cornered vehicles and close coupling of tractor units and trailers can reduce drag. Wind deflectors fitted to cab roofs can, if well designed and matched to the vehicle, give useful savings. These depend greatly on operating conditions, but a 10% saving is a reasonable expectation. In high cross-winds the effectiveness of deflectors may be much reduced.

11/2 The fitting of correctly-designed caps or deflectors to lorry cabs can streamline air flow round vehicles and effect useful savings, especially with operation at high speed on motorways.
Weetabix Ltd, Burton Latimer, Northamptonshire.

As well as reducing fuel consumption at the same speed, the fitting of a deflector usually increases the vehicle's top speed, so that potential savings may not materialise if the driver does not limit his speed. Speed-control devices are, therefore, often fitted in conjunction with air deflectors.

The large fleet operator has good opportunities to conduct trials using various types of deflector and other devices on similar vehicles in order to obtain reliable comparative costs. Some independent bodies like the Society of Automobile Engineers (SAE) in the USA and the Motor Industry Research Association (MIRA) in the UK, have undertaken studies under controlled conditions and their reports are of particular value, for example MIRA report

K23049, 'Assessment of proprietary devices for reducing the aerodynamic drag of commercial vehicles'.

The correct inflation of tyres has been referred to earlier. The selection of tyres is also important. Low-profile radial tyres can give a 5% fuel saving over cross-ply tyres because of their lower rolling resistance. For petrol (gasoline) engines, electronic ignition overcomes many of the shortcomings of traditional ignition systems and generally aids fuel economy.

11.4 Vehicle maintenance

Regular skilled maintenance is an essential part of any effort to cut transport fuel costs. Vehicle manufacturers' schedules should be adhered to and a separate and complete record kept for each vehicle. In considering maintenance requirements due regard should be paid to the conditions of use of the vehicles, including the materials carried, skill and experience of the drivers, surfaces over which they operate as well as the more obvious criteria of mileage and hours of use.

In the course of maintenance, particular attention should be paid to those aspects that affect fuel consumption:

Diesel engine pumps and injectors.

Petrol (gasoline) engine mixture, timing, idling speed, points and plugs.

Air and fuel filters on all types of engine.

Engine thermostat settings.

Fuel tanks and lines, to ensure that there are no leaks.

Wheel and chassis alignment.

Engine speed governors and throttle stops, when fitted.

Brakes, to ensure that they are not binding.

Re-jetting carburettors and resetting diesel-fuel injection rates has been shown to produce savings of up to 10%. Carburettors designed for economy are available but many gimmicky devices claimed to save fuel have proved of little practical value.

Responsibility for the maintenance of all forms of transport should preferably be vested in one person who should also control the training of drivers and mechanics. Where a properly equipped workshop and skilled mechanics cannot be justified, maintenance should be left to a dealer or competent contractor who does have the facilities. Vehicle maintenance is not a suitable task for the odd-job man.

Correct grades of lubricating oils and greases should be used and steps taken to avoid contamination by water or abrasive matter. The viscosity of the oil used for engines, particularly diesel engines, must be matched to the type, operating conditions and climate in which they have to work. Clear instructions on routine lubrication (e.g. self-adhesive guides fixed inside the cab) should be available for long-distance drivers, especially those involved in intercontinental haulage.

Vehicles should be garaged under cover at night, especially in winter, to improve cold starting and avoid engine revving to defrost the cab. The overfilling of vehicle tanks during refuelling, which is dangerous as well as wasteful, should be strongly discouraged. Nozzle cut-offs should be fitted. The correct grade of fuel to be used should be indicated on the vehicle tank, and petrol (gasoline)-engined vehicles should not use a fuel of higher octane rating than necessary.

- **Ensure that every vehicle is serviced regularly to a predetermined schedule and that individual vehicle records are maintained. In compiling schedules, consider exactly how each vehicle is used.**
- **Ensure that mechanics are properly trained and that they pay** *particular* **attention to items that affect economic operation.**
- **Use correct grades of fuels, lubricants and greases.**
- **Garage all vehicles under cover, if possible.**
- **Maintain proper control over refuelling.**

APPENDIX A Units and conversion factors

SI units

Full information about the International System of units is given in a publication from the International Bureau of Weights and Measures, *Le Système International d'Unités.* Authorised English translations have been published in the United Kingdom by HMSO and in the United States of America by the NBS.

Although international agreement exists on the base units, universal agreement has not yet been reached on all other units and multiples, and some non-SI units already in common use may be retained. A number of countries have now published their own SI standards or guiding notes which should be consulted.

Base units

Quantity	*Name*	*Symbol*
Length	metre	m
Mass	kilogram	kg
Time	second	s
Electric current	ampere	A
Temperature	kelvin	K
Luminous intensity	candela	cd
Amount of substance	mole	mol

Other SI supplementary, derived and related units

Quantity	*Name*	*Symbol*	
Plane angle	radian	rad	
Solid angle	steradian	sr	
Area	hectare	ha	(= 10^4 m^2)
Mass	tonne	t	(= 10^3 kg)
Frequency	hertz	Hz	(= 1 s^{-1})
Force	newton	N	(= 1 kg m/s^2)
Pressure	pascal	Pa	(= 1 N/m^2)
Pressure	bar	bar	(= 10^5 Pa)
Temperature	degree Celsius	°C	(= K – 273.15)
Energy	joule	J	(= 1 N m)

Power	watt	W	(= 1 J/s)
Luminous flux	lumen	lm	(= 1 cd/sr)
Illuminance	lux	lx	(= 1 lm/m^2)
Kinematic viscosity	stokes	St	(= 10^{-4} m^2/s)
Dynamic viscosity	poise	P	(= 10^{-1} Pa s)
Volume	litre	l	(= 1 dm^3)

Electrical

Quantity	coulomb	C	(= 1 A s)
Potential	volt	V	(= 1 W/A)
Resistance	ohm	Ω	(= 1 V/A)
Capacitance	farad	F	(= 1 A s/V)
Inductance	henry	H	(= 1 V s/A)

Multiples and sub-multiples

Factor	*Prefix*	*Symbol*
10^{12}	tera	T
10^{9}	giga	G
10^{6}	mega	M
10^{3}	kilo	k
10^{2}	hecto	h
10^{1}	deca	da
10^{-1}	deci	d
10^{-2}	centi	c
10^{-3}	milli	m
10^{-6}	micro	μ
10^{-9}	nano	n
10^{-12}	pico	p
10^{-15}	femto	f
10^{-18}	atto	a

Approximate conversion factors for various units in common use

Length

1 cm	=	0.394 in	1 in	=	25.4 mm
1 m	=	3.281 ft	1 ft	=	30.48 cm
1 m	=	1.094 yd	1 yd	=	0.9144 m
1 km	=	0.621 mile	1 mile	=	1.609 km
1 mm	=	10^3 μm			

Area

1 cm^2	=	0.155 in^2	1 in^2	=	6.452 cm^2
1 m^2	=	10.76 ft^2	1 ft^2	=	0.093 m^2
1 km^2	=	0.386 $mile^2$	1 yd^2	=	0.836 m^2
1 ha	=	2.471 acre	1 $mile^2$	=	2.590 km^2
			1 acre	=	0.405 ha

Volume, capacity

1 m^3	=	35.31 ft^3	1 in^3	=	16.39 cm^3
1 m^3	=	220 Imp gal	1 ft^3	=	0.0283 m^3
1 m^3	=	264 US gal	1 ft^3	=	28.32 litre
1 m^3	=	6.29 barrel	1 ft^3	=	6.23 Imp gal
1 litre	=	0.22 Imp gal	1 ft^3	=	7.48 US gal
1 litre	=	0.264 US gal	1 pint	=	0.568 litre
			1 barrel	=	42 US gal
			1 barrel	=	159 litre

Velocity, flow rate

1 m/s	=	2.24 mph	1 ft/s (fps)	=	0.305 m/s
1 m/s	=	3.28 ft/s (fps)	1 mph	=	0.447 m/s
1 km/h	=	27.78 cm/s	1 mph	=	1.61 km/h
1 km/h	=	0.911 ft/s (fps)	1 ft^3/min	=	0.472 l/s
1 l/s	=	2.119 ft^3/min	1 ft^3/s	=	0.028 m^3/s
1 m^3/h	=	3.67 Imp gal/min	1 Imp gal/min	=	0.273 m^3/h

Mass

1 kg	=	35.27 ounce	1 ounce	=	28.35 g
1 kg	=	2.205 lb	1 lb	=	0.4536 kg
1 tonne	=	2205 lb	1 cwt	=	50.8 kg
1 tonne	=	0.984 Imp ton	1 Imp ton	=	1016 kg
1 tonne	=	1.102 US ton	1 Imp ton	=	1.12 US ton
			1 US ton	=	907 kg
			1 grain	=	64.8 mg

Density, concentration

1 g/m^3	=	0.448 $grains/ft^3$	1 lb/in^3	=	27.68 Mg/m^3
1 kg/m^3	=	0.0624 lb/ft^3	1 lb/ft^3	=	16.02 kg/m^3

1 kg/m^3	=	0.1 lb/Imp gal	1 lb/ft^3	=	0.1605 lb/Imp gal
1 kg/m^3	=	0.835 lb/US gal	1 lb/ft^3	=	0.134 lb/US gal
			1 lb/Imp gal	=	99.8 kg/m^3
			1 lb/US gal	=	119.8 kg/m^3

Pressure

1 Pa	=	1 N/m^2	1 atm	=	760 mmHg
1 Pa	=	0.004 inH_2O	1 atm	=	14.696 lbf/in^2
1 kPa	=	0.145 lbf/in^2 (psi)	1 atm	=	101.3 kPa
1 kPa	=	0.295 inHg	1 bar	=	100 kPa
1 kPa	=	7.5 mmHg	1 lbf/in^2 (psi)	=	6.895 kPa
1 kgf/cm^2	=	14.22 lbf/in^2 (psi)	1 inH_2O	=	249 Pa
			1 inHg	=	3386 Pa
			1 torr (mmHg)	=	133.3 Pa

Force

1 N	=	10^5 dyne	1 lbf	=	4.448 N
1 N	=	0.102 kgf	1 Imp tonf	=	9.964 kN
1 N	=	0.2248 lbf			
1 kgf	=	9.81 N			

Energy, work, heat

1 J	=	0.239 cal	1 cal	=	4.187 J
1 J	=	0.738 ft lbf	1 Btu	=	1.055 kJ
1 J	=	10^7 ergs	1 Btu	=	0.293 Wh
1 kJ	=	0.948 Btu	1 therm	=	10^5 Btu
1 MJ	=	0.0095 therm	1 therm	=	105.5 MJ
1 MJ	=	0.3725 hp hour	1 therm	=	29.31 kWh
1 MJ	=	0.278 kWh	1 ft lbf	=	1.356 J
1 kWh	=	3.60 MJ	1 hp hour	=	2545 Btu
			1 hp hour	=	2.685 MJ

Power, heat flow rate

1 W	=	0.86 kcal/h	1 Btu/h	=	0.293 W
1 W	=	44.25 ft lbf/min	1 kcal/h	=	1.163 W
1 kW	=	3412 Btu/h	1 hp	=	0.746 kW
1 kW	=	1.341 hp	1 hp	=	42.42 Btu/min
1 kW	=	56.87 Btu/min	1 hp	=	33 000 ft lbf/min

Calorific value, heat content

1 MJ/kg	=	238.8 kcal/kg	1 Btu/lb	=	0.5556 cal/g
1 MJ/kg	=	430 Btu/lb	1 Btu/lb	=	2.326 kJ/kg
1 MJ/m^3	=	26.84 Btu/ft^3	1 Btu/ft^3	=	37.26 kJ/m^3
1 cal/g	=	4.187 kJ/kg	1 Btu/Imp gal	=	232.1 kJ/m^3
			1 Btu/US gal	=	278.7 kJ/m^3

Heat flux

1 W/m^2	=	0.317 Btu/ft^2h	1 Btu/ft^2h	=	3.155 W/m^2
1 kW/m^2	=	860 $kcal/m^2h$	1 $kcal/m^2h$	=	1.163 W/m^2

Combustion intensity

1 kW/m^3	=	96.6 Btu/ft^3h	1 Btu/ft^3h	=	10.35 W/m^3
1 kW/m^3	=	860 $kcal/m^3h$	1 $kcal/m^3h$	=	1.163 W/m^3

Heat transfer coefficient

1 W/m^2 °C	=	0.176 Btu/ft^2 h °F	1 Btu/ft^2 h °F	=	5.678 W/m^2 °C
1 W/m^2 °C	=	0.86 $kcal/m^2$ h °C	1 $kcal/m^2$ h °C	=	1.163 W/m^2 °C

Thermal conductivity

1 W/m °C	=	0.578 Btu/ft h °F	1 Btu/ft^2 h °F	=	1.73 W/m °C
1 W/m °C	=	6.93 Btu in/ft^2 h °F	1 Btu in/ft^2 h °F	=	0.144 W/m °C

Specific heat capacity

1 kJ/kg °C	=	0.239 Btu/lb °F	1 Btu/lb °F	=	4.187 kJ/kg °C
1 kJ/m^3 °C	=	0.0149 Btu/ft^3 °F	1 Btu/ft^3 °F	=	67.07 kJ/m^3 °C

Temperature

Absolute zero (0 kelvin) = –273.15°C = –459.8°F

$$°C = \frac{5}{9}(°F - 32) \qquad °F = \frac{9}{5}°C + 32$$

Temperature intervals

1 deg C = 1.8 deg F

1 deg F = 0.556 deg C

Calorific values

Typical gross calorific values (or higher heating values) are given below for a number of fuels. More precise values should always be obtained, if possible, for particular fuels being used. Suppliers will usually provide data.

Gross CV represents the total heat content of the fuel on combustion, including the heat that would be obtained by condensing water vapour in the flue gases. Net CV is the gross value less this latent heat of vaporisation. The difference between gross and net values thus depends upon the hydrogen content of the fuel.

Except for very approximate figures, the conditions under which CV is determined should always be stated. Gases are measured at standard temperatures and pressures, dry or saturated with water vapour. Solid fuels contain varying amounts of water and mineral matter (most of which becomes ash). Analysis and CV determination must be on some uniform basis — usually referred to as dry, ash-free (daf) or dry, mineral matter free (dmmf).

The condition of fuels as delivered and used may be very different from the standard data, especially for solid fuels. A knowledge of a coal's rank, free moisture, ash content and size are essential in considering its suitability for any particular application, and for determining performance.

Typical gross calorific values

	(MJ/m^3)	(Btu/ft^3)	(MJ/kg)	(Btu/lb)
Fuel gases (dry; 60°F, 30 inHg)				
Producer gas (cold)	5.0	135		
Blast furnace gas	3.5	95		
Coke-oven gas	19.5	525		
Natural gas	37.3	1 000		
Acetylene	55.9	1 500	49.9	21 450
Commercial butane	121.8	3 270	49.3	21 200
Commercial propane	93.1	2 500	50.0	21 500

	(MJ/l)	(Th/gal)	(MJ/kg)	(Btu/lb)
Liquid fuels (water-free)				
Kerosine	36.7	1.58	46.5	20 000
Gas oil (diesel fuel)	38.0	1.64	45.6	19 600
Light fuel oil	40.1	1.73	43.5	18 700
Medium fuel oil	40.4	1.74	43.1	18 500
Heavy fuel oil	40.8	1.76	42.6	18 300
Light CTF (50/100)	39.4	1.70	39.6	17 000
Heavy CTF (300)	45.3	1.95	37.9	16 300

	(MJ/kg)	(Btu/lb)
Solid fuels (dry, ash-free basis)		
Lignite	26.7	11 500
High volatile coal	33.7	14 500
Low volatile coal	36.5	15 700
Anthracite	36.3	15 600
Coke breeze	25.6	11 000
Large coke	29.1	12 500
Wood	19.8	8 500
Peat	22.1	9 500
Household refuse	11.6	5 000

APPENDIX B Test equipment

Expenditure on good-quality, permanent instrumentation and controls can be justified on larger items of energy-using plant, especially steam boilers and high-temperature furnaces. For smaller applications, for spot checks and investigations, and in situations where fixed instruments would be vulnerable to damage, portable test equipment and easily refitted meters should be used.

The examples illustrated are simple to operate and, with reasonable care, robust enough for industrial use. Operators need only elementary training before being able to obtain useful results with them.

In most cases, similar equipment is available from other suppliers, but the items shows are considered to be reliable and are made or distributed by reputable organizations.

Universal test plug

The taking of temperature, pressure and flow readings in pipes and ducts is facilitated by fitting permanent self-sealing test plugs. Simple dial instruments (as illustrated) or other portable instruments fitted with suitable probes may be used. The core — suitable for most gases and liquids up to 40 bar and 200°C — closes when the probe is removed. Useful where permanent gauges would be vulnerable to damage.

Binder Engineering Co Ltd, Banbury, Oxfordshire.

Portable electronic thermometer
An adaptable temperature measurement tool. The battery-powered basic instrument (L) when housed in a carrying case (R) is suitable for laboratory or industrial use. The model shown covers the temperature range 0–600° (in 100°C bands with 1°C accuracy) and uses various plug-in thermocouple probes. By using a balanced switchbox in conjunction with the instrument up to six points can be monitored in rapid succession.
Polkinghorne Industries Ltd, Workington, Cumbria

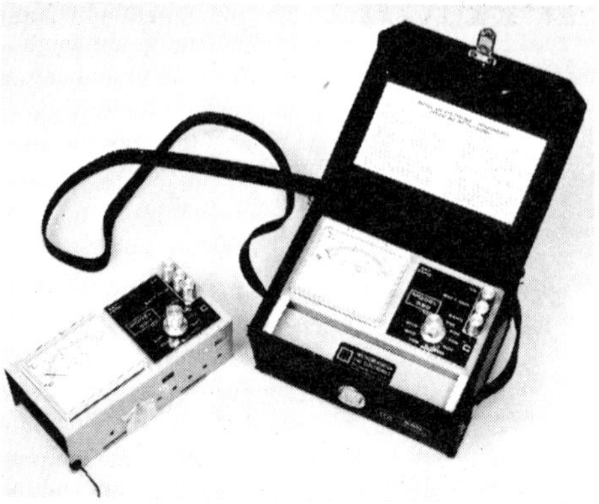

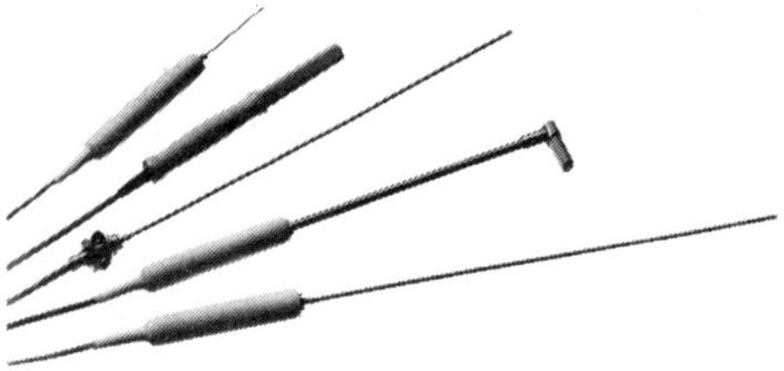

Thermocouple probes
No matter what sort of indicating instrument is employed, the thermocouple used should be carefully selected to match the application and properly positioned if a representative temperature is to be measured. The same care is needed for all sensing devices — thermocouples, bimetals, resistance elements, fluid expansion and vapour pressure bulbs. The illustration shows thermocouple probes suitable for a variety of different applications.
Comark Electronics Ltd, Rustington, West Sussex.

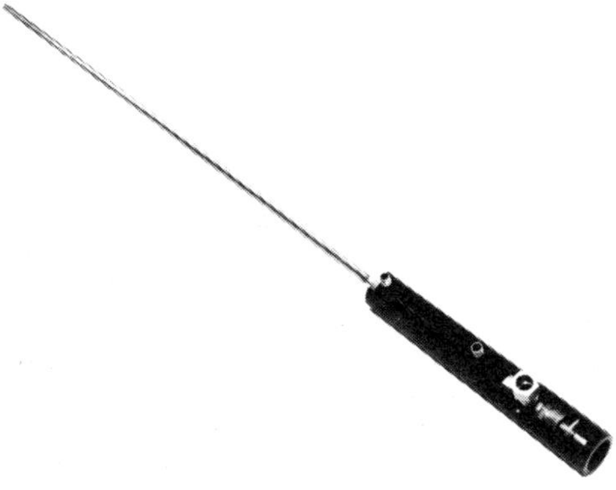

Suction pyrometer
Errors arise if a normal sheathed thermocouple is used to measure gas temperatures, especially high ones. The suction pyrometer overcomes these by shielding the thermocouple from wall radiation and drawing gases over it at high velocity to ensure good convective heat transfer. The thermocouple thus produces a reading which approaches the true temperature at the sampling point rather than a temperature between that of the walls and the gases.
Land Combustion Ltd, Sheffield, South Yorkshire.

Digital thermometer
A pocket-sized battery-operated thermometer is especially convenient for spot checks or where a number of rapid readings of process temperatures need to be taken. The model illustrated, using NiCr/NiAl thermocouples, has an accuracy of 2°C between −50°C and +800°C, and will operate up to 1000°C. The bright phosphorescent digital display may be held while the reading is noted.
Comark Electronics Ltd, Rustington, West Sussex.

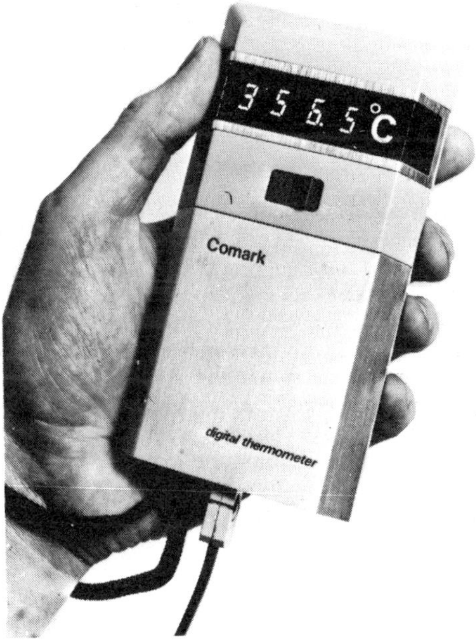

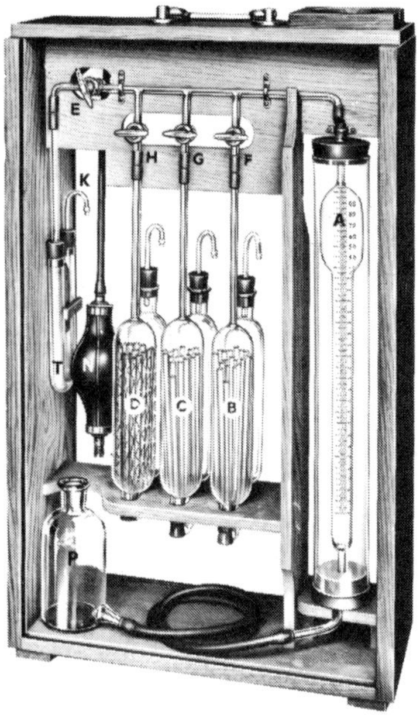

Orsat gas analyser
The standard apparatus for determining carbon dioxide, oxygen and carbon monoxide in flue gases. The Orsat should be used, in preference to simple indicators, where higher accuracy is required (e.g. for calibrating instruments). A little skill is required to use and maintain an Orsat properly and because the working parts are glass it must also be handled with reasonable care (see BS 1756 Part 2).
Shawcity Ltd, Faringdon, Oxfordshire.

Combustion indicators
The accuracy of the Orsat appartus is unnecessary for many rapid on-the-spot measurements of CO_2 and O_2 levels, which can more conveniently be carried out with these indicators. They are simple to use and a complete combustion testing kit can be carried from place to place without dismantling or draining the instruments. Various combinations of indicators, thermometers, etc., are available as portable kits.
Shawcity Ltd, Faringdon, Oxfordshire.

Multi-gas detector
The detector consists of a prepared disposable tube inserted into a bellows pump. The pump draws 100 cm^3 at each stroke and a measured sample is passed through the glass detector tube. The concentration of the gas being investigated is indicated directly by the length of discolouration. Different tubes are available for measuring the concentration of about 100 gases and vapours.
Draeger Safety Ltd, Chesham, Buckinghamshire.

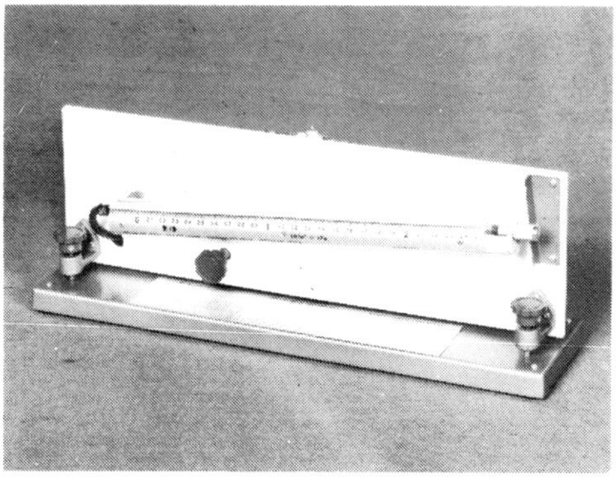

Portable manometer
A manometer (or draught gauge) may be used to measure positive or negative pressures (e.g. furnace pressures) or, in combination with a suitable pitot tube or orifice plate, differential pressures, and hence flow rates in pipes and ducts. The portable manometer illustrated can readily be levelled and zeroed. Care is needed in making differential pressure measurements, and a recognised procedure should be used which will give meaningful results.
Airflow Developments Ltd, High Wycombe, Buckinghamshire.

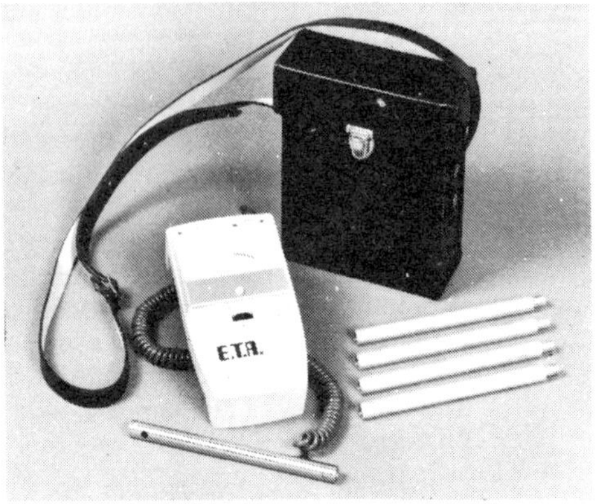

Air velocity meter
A compact, battery-operated, hand-held anemometer capable of measuring air velocity between 0.1 m/s and 15.0 m/s (3000 ft/min), with high accuracy at the lower end of the scale. The probe, which is compensated for air temperature, may be fitted with extension rods if required. Useful for measuring air velocity from grilles in air conditioning systems or air movement in rooms. Not suitable for high temperatures.
Airflow Developments, Ltd,
High Wycombe, Buckinghamshire.

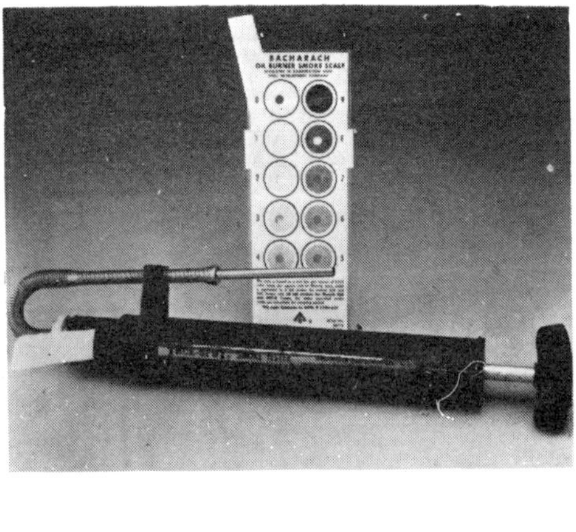

Smoke tester
Smoke is unburned carbon which fouls heat exchange surfaces, wastes fuel and pollutes the air. The smoke tester is a simple positive method of checking smoke content. A measured volume of flue gas is drawn through a filter paper with the pump. The resulting smoke spot is compared visually with a standard scale to give a grade of smoke density, and hence an indication of completeness of combustion.
Shawcity Ltd, Faringdon,
Oxfordshire

Dewpoint meter

The equipment illustrated is designed to measure the sulphuric acid dewpoint of combustion gases inside ovens, economisers, flues, etc. A detector probe is inserted which is progressively cooled with air until the onset of condensation is indicated. A useful tool to assess plant performance and to monitor measures taken to eliminate corrosion and fouling in industrial plant, which are counter-productive to efficient operation.

Land Combustion Ltd, Sheffield, South Yorkshire.

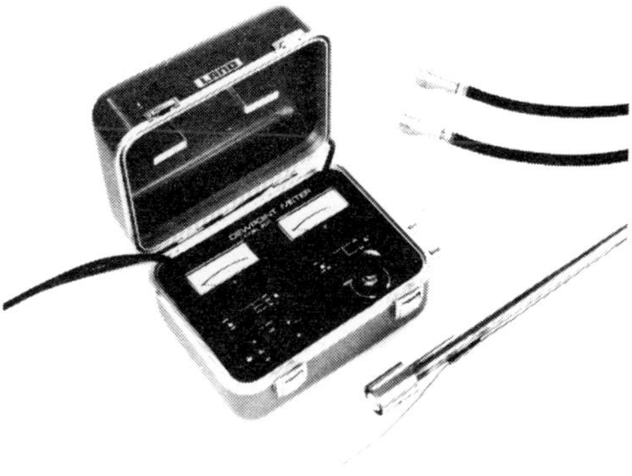

Oil meter

With an accuracy of $\pm 1\%$, simple, positive-displacement oil meters are invaluable for measuring the consumption of individual boilers and other oil-fired plant. Many other non-corrosive liquids can also be measured with this type of meter which has a simple integrating register fitted with a protective hinged cover. With union connections, meters can be easily inserted in prepared pipelines for spot checks.

Arkon Instruments Ltd, Cheltenham, Gloucestershire.

Gas Meter
A rotary meter designed to provide accurate measurement even at low flow rates. Suitable for primary or distribution metering. The small rotating pointer is useful for spot measurements of gas flow rates, by the timing of a complete revolution.
International Gas Apparatus, Camberley, Surrey

Industrial light meter
As well as being self-contained and robust, for use in an industrial environment, a light meter should also be capable of accepting light from sources having low angles of incidence. The meter illustrated will directly measure illumination from tungsten filament and fluorescent sources, and daylight. Multipliers are provided to measure levels above 500 lux. The pointer may be locked while a reading is noted.
Sangamo Weston Ltd, Enfield, Middlesex.

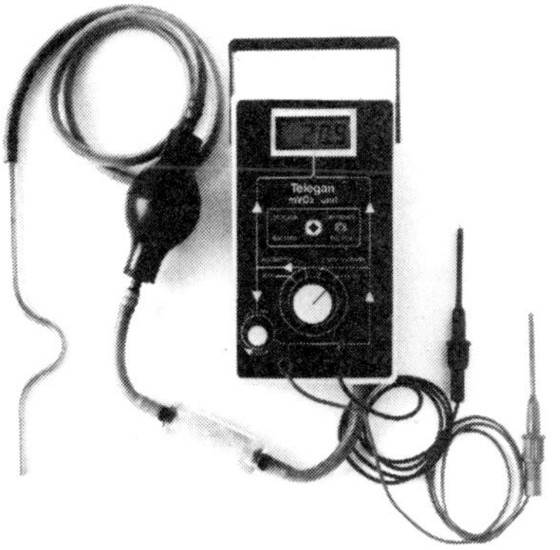

Portable oxygen analyser
A hand-held instrument with liquid crystal display to indicate the oxygen content of a sample of flue gases drawn from a duct or chimney by means of the hand-bulb pump supplied with it. The instrument may also be used as a millivolt source or as a millivolt meter.
Telegan Ltd, Croydon, Surrey CR9 2BU.

Steam trap indicator
A failure indicator, suitable for all types of steam trap, which shows clearly by means of a red or green light, whether or not a trap under test is passing steam. The hand-held indicator is easily plugged in to sensor chambers permanently fitted upstream of the traps they serve.
Spirax Sarco Ltd, Cheltenham, Gloucestershire.

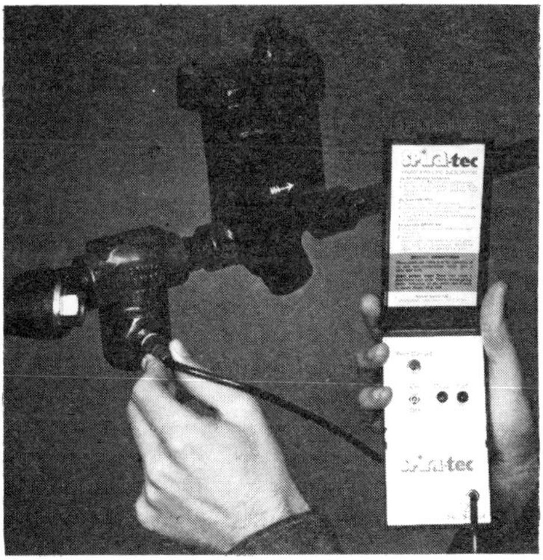

APPENDIX C Further reading and reference

A selection of publications covering a number of topics dealt with in this handbook is given below. The list is by no means comprehensive and is intended only as a guide to further reading and reference. Though now out of date in some respects, several older works have been included because they deal particularly well with the principles of efficient energy utilisation.

A number of British Standards of relevance are listed separately. Similar standards are available in many other countries, together with ISO standards. Standards handbooks and yearbooks are a fruitful source of information on fuels, energy-using equipment and test methods.

Energy management and conservation

Handbook of Industrial Energy Analysis
I. Boustead and G. F. Hancock
Ellis Horwood, Chichester, UK, 1979

Industrial Energy Conservation
M. H. Chiogioji
Marcel Dekker, New York, 1979

Energy Conservation in Distribution
P. N. C. Cooke (ed)
Gower Press, Farnborough, UK, 1980

Cutting Energy Costs
R. Dick-Larkam
Gower Press, Farnborough, UK, 1977

Industrial Energy Conservation (2nd edition)
D. A. Reay
Pergamon, Oxford, UK, 1979

Energy Conservation and Energy Management in Buildings
A. F. C. Sherratt
Applied Science Publishers, Barking, Essex, UK, 1976

Fuel Economy Handbook (2nd edition)
W. Short (ed)
Graham & Trotman, London, 1979

Industrial Energy Management for Cost Reduction
T. E. Smith
Ann Arbor Science, Michigan, USA (John Wiley, Chichester, UK), 1979

Plant Engineers' and Managers' Guide to Energy Conservation
A. Thumann
Van Nostrand Reinhold, New York, 1977

Fuels production and data

An Introdution to Coal Technology
N. Berkowitz
Academic Press, New York

Gas Making and Natural Gas
British Petroleum, London, 1972

Hydrocarbon Fuels
E. M. Goodger
Macmillan, London, 1975

Modern Petroleum Technology (4th edition)
G. D. Hobson and W. Pohl
Applied Science Publishers, Barking, Essex, UK, 1973

Coal Tar Fuels
W. H. Huxtable
Association of Tar Distillers, London, 1961

Substitute Natural Gas: Manufacture and Properties
W. L. Lom and A. F. Williams
Applied Science Publishers, Barking, Essex, UK, 1976

Technical Data on Fuel (7th edition, SI units)
J. W. Rose and J. R. Cooper (eds)
World Energy Conference (BNC), London, 1976

The Petroleum Handbook (5th edition)
Shell International Petroleum, London, 1966

Fuel science and technology

Temperature Measurement in Engineering (2 volumes)
H. D. Baker, E. A. Ryder and N. H. Baker
Omega Press, London

Energy Technology Handbook
D. M. Considine (ed)
McGraw-Hill, New York, 1977

The Efficient Use of Energy
I. G. C. Dryden (ed)
IPC Science and Technology Press, Guildford, UK, 1975

Fuels and Fuel Technology (2 volumes)
W. H. Francis
Pergamon, Oxford, UK, 1965

Fuel Science
J. H. Harker and D. A. Allen
Oliver & Boyd, London, 1972

Radiative Transfer
H. C. Hottel and A. F. Sarofim
McGraw-Hill, New York, 1967

Energy Storage
J. Jensen
Newnes-Butterworths, London, 1979

Fuels and Combustion Handbook
A. E. Johnson and G. H. Auth (eds)
McGraw-Hill, New York, 1951

Heat Transmission (3rd edition)
W. H. McAdams
McGraw-Hill, New York, 1954

Introduction to the Study of Fuel
J. C. Macrae
Elsevier, Amsterdam, 1966

Industrial Gas Utilisation: Engineering Principles and Practice
R. Pritchard, J. J. Guy and N. E. Connor
Bowker, London, 1977

Engineering Thermodynamics, Work and Heat Transfer (SI and British units editions)
G. F. C. Rogers and Y. R. Mayhew
Longman, London, 1967

Engineering reference

Standard Handbook for Mechanical Engineers (7th edition)
T. Baumeister and L. S. Marks
McGraw-Hill, New York, 1967

McGraw-Hill Encyclopedia of Energy
D. N. Lapedes
McGraw-Hill, New York, 1976

Chemical Engineers' Handbook (5th edition)
R. H. Perry and C. H. Chilton (eds)
McGraw-Hill, New York, 1973

Gas Engineers' Handbook
C. G. Segeler
Industrial Press, New York, 1966

Instrumentation and control

Process Instruments and Controls Handbook
D. M. Considine
McGraw-Hill, New York, 1974

Instrument Technology (3 volumes)
E. B. Jones
Newnes-Butterworths, London

Instrumentation in Industry
H. E. Soisson
John Wiley, New York, 1975

Instrumentation in Process Control
E. J. Wightman
Butterworth, London, 1972

Boilers and steam production

Steam — Its Generation and Use
(39th edition)
Babcock & Wilcox, New York, 1978

Fundamentals of Boilerhouse Technique
P. D. Dehnel
Hutchinson, London, 1973

Boiler House Design for Solid Fuel
National Coal Board, London, 1975

Boiler House Practice
J. N. Williams
Allen & Unwin, London, 1969

Steam Plant Operation (4th edition)
E. B. Woodruff and H. B. Lammers
McGraw-Hill, New York, 1977

Steam properties and utilisation

Thermal Efficiency and Power Production
K. Fenton
Pitman, London, 1966

Steam Storage Installations
N. Goldstern
Pergamon, Oxford, UK, 1970

The Efficient Use of Steam
P. M. Goodall (ed)
IPC Science and Technology Press, Guildford, UK, 1980

Enthalpy–Entropy Diagram for Steam (SI units)
D. C. Hickson and F. R. Taylor
Blackwell, Oxford, UK, 1966

Thermodynamic and Transport Properties of Fluids (SI units)
Y. R. Mayhew and G. F. C. Rogers
Blackwell, Oxford, UK, 1967

Steam Trapping and Air Venting
(5th edition, SI units)
L. G. Northcroft and W. M. Barker
Hutchinson, London, 1979

Industrial space heating and air conditioning

Handbook of Fundamentals
American Society of Heating, Refrigeration and Air Conditioning Engineers
New York

Building for Energy Conservation
P. Burberry
Architectural Press, London, 1978

The CIBS Guide (sections)
The Chartered Institution of Building Services, London

Woods' Practical Guide to Fan Engineering
B. B. Daly
Woods of Colchester, Colchester, UK, 1978

Heating and Air-Conditioning of Buildings (6th edition, SI units)
J. R. Kell and P. L. Martin
Architectural Press, London, 1978

Principles of Air Conditioning (3rd edition)
P. V. Land
Delmar, New York, 1979

Handbook of Heating, Ventilating and Air Conditioning (7th edition)
J. Porges and F. Porges
Newnes-Butterworths, London, 1976

The Trane Air Conditioning Manual
Trane Publications, Sunbury-on-Thames, UK

Furnaces and process heating

Calculations in Furnace Technology
C. Davies
Pergamon, Oxford, UK, 1970

Drying: Principles and Practice
R. B. Keey
Pergamon, Oxford, UK, 1972

Introduction to Industrial Drying Operations
R. B. Keey
Pergamon, Oxford, UK, 1978

The Science of Flames and Furnaces
M. W. Thring
Chapman & Hall, London, 1952

Industrial Furnaces (2 volumes)
W. Trinks and M. H. Mawhinney
John Wiley, New York; Chapman & Hall, London, 1961, 1967

Insulation and refractories

Refractories: Production and Properties
J. H. Chesters
Iron and Steel Institute, London, 1973

The Insulation Handbook (annual)
Comprint, Watford, UK

Thermal Insulation of Buildings
C. C. Handisyde and D. J. Melluish
HMSO, London, 1971

Thermal Insulation
J. F. Malloy
Van Nostrand Reinhold, New York, 1969

Refractories (4th edition)
F. H. Norton
McGraw-Hill, New York, 1968

Refractories and their Uses
K. Shaw
Applied Science Publishers, London, 1972

Waste handling and recovery

Practical Heat Recovery
J. L. Boyen
John Wiley, New York, 1976

Heat Pipes (2nd edition)
P. D. Dunn and D. A. Reay
Pergamon, Oxford, UK, 1978

Heat Pumps
R. D. Heap
E. & F. N. Spon, London, 1979

Waste Heat Recovery Conference
The Institute of Fuel
Chapman & Hall, London, 1963

A Guide to Chimney Design
K. R. Jackson
IPC Science and Technology Press, Guildford, UK, 1978

Waste Management, Control, Recovery and Re-use
N. Y. Kirov
Ann Arbor Science, Michigan, USA, 1975

The Heat Pump (bibliography)
S. Loyd and C. Starling (eds)
BSIRA, Bracknell, UK, 1975

Heat Recovery from Buildings (bibliography)
S. Loyd and C. Starling (eds)
BSIRA, Bracknell, UK, 1976

Heat Recovery Systems
D. A. Reay
E. & F. N. Spon, London, 1979

Heat Pumps. Design, Construction and Applications
D. A. Reay and B. Macmichael
Pergamon, Oxford, UK, 1979

Non-fuel energy sources

Geothermal Energy
H. C. H. Armstead
E. & F. N. Spon, London, 1978

The Generation of Electricity by Windpower (2nd edition)
E. W. Golding
E. & F. N. Spon, London, 1976

Heat and Power from the Sun (3rd edition) (bibliography)
S. Loyd and C. Starling (eds)
BSIRA, Bracknell, UK, 1977

Sun Power
J. C. McVeigh
Pergamon, Oxford, UK, 1977

Biological Energy Resources
M. Slesser and C. Lewis
E. & F. N. Spon, London, 1979

Solar Energy
J. R. Williams
John Wiley, New York, 1977

Solar Energy
J. I. B. Wilson
Wykeham, London, 1979

Electricity distribution and use

IES Code for Interior Lighting (Section B9 of CIBS Guide)
Chartered Institution of Building Services, London

Principles of Electrical Technology
H. Cotton
Pitman, London, 1967

Induction Heating Handbook
J. Davies and P. Simpson
McGraw-Hill, New York

Lamps and Lighting
S. T. Henderson and A. M. Marsden
Arnold, London, 1972

IES Lighting Handbook
Illuminating Engineering Society, New York

Management Guide to Modern Industrial Lighting
S. L. Lyons
Applied Science Publishers, London, 1972

The Electrical Engineer's Reference Book (13th edition)
M. G. Say
Newnes-Butterworths, London, 1976

Power System Economics
E. O. Taylor and G. A. Boal
Arnold, London, 1969

Reducing the Cost of Electricity Supply
A. S. Windett
Gower Press, London, 1973

Engines and on-site power generation

Total Energy
R. M. E. Diamant
Pergamon, Oxford, UK, 1970

Total Energy Conference
Vol 1 — papers, Vol 2 — discussion
The Institute of Fuel, London, 1971

Total Energy — The On-Site Production of Heat and Power (bibliography)
S. Loyd (ed)
BSIRA, Bracknell, UK, 1975

Internal Combustion Engines
E. F. Obert
International Textbook, London, 1968

Selection and Use of Standby and Peak Lopping Generating Equipment
(Paper 362)
A. B. Shearer
Diesel Engineers' and Users' Association, London, 1974

Pneumatics and hydraulics

Guide to the Selection and Installation of Compressed Air Services (2nd edition)
British Compressed Air Society, London, 1979

Applied Hydraulics and Pneumatics in Industry
J. R. Fawcett
Trade and Technical Press, Morden, UK, 1968

Pneumatic Handbook (5th edition)
R. H. Warring (ed)
Trade and Technical Press, Morden, UK, 1978

Industrial Air Compressors
F. G. White
Foulis, London, 1967

Industrial water

Water Conditioning for Industry
S. T. Powell
McGraw-Hill, New York, 1954

Cooling Towers. Principles and Practice
(2nd edition)
W. Stanford and G. B. Hill
Carter Industrial Products Limited, Birmingham, UK, 1972

Industrial lubrication

Industrial Lubrication
M. G. Billett
Pergamon, Oxford, UK, 1979

Lubrication and Lubricants
E. R. Braithwaite (ed)
Elsevier, Amsterdam, 1967

Lubrication Theory and Its Application
BP Trading Limited, London, 1969

Lubrication in Practice
G. G. Evans (ed)
Macmillan-Esso, Basingstoke, UK, 1972

Many other publications provide useful guidance on industrial energy conservation, such as those issued by:

Government departments. Department of Energy, Department of the Environment, Department of Industry etc.

Professional bodies. Institute of Energy, Watt Committee on Energy, Chartered Institution of Building Services etc.

Energy suppliers. National Coal Board, British Gas, Electricity Council, Shell, BP, Esso etc.

Trade and technical associations. British Combustion Equipment Manufacturers' Association, Liquefied Petroleum Gas Industry Technical Association etc.

Equipment and materials suppliers. Spirax-Sarco, Cape Insulation, APV etc.

British Standards

Summaries of the contents of *all* standards and codes of practice are given in the current *British Standards Yearbook.*

BS 350:(parts)	Conversion factors and tables.
BS 526:1961	Definitions of the calorific value of fuels.
BS 799:(parts)	Specification for oil-burning equipment.
BS 845:1972	Acceptance tests for industrial type boilers and steam generators.
BS 1041:(parts)	Code for temperature measurement.
BS 1042:(parts)	Methods for the measurement of fluid flow in pipes.
BS 1374:1972	Recommendations on the use of British Standard log sheets for steam and hot water boiler plants.
BS 1469:1962	Coal tar fuels.
BS 1710:1975	Identification of pipelines.
BS 1756:(parts)	Methods for the sampling and analysis of flue gases.
BS 2869:1970	Petroleum fuels for oil engines and burners. (In course of revision.)
BS 2885:1974	Code for acceptance tests on stationary steam generators of the power station type.
BS 3048:1958	Code for the continuous sampling and automatic analysis of flue gases. Indicators and recorders.
BS 3812:1964	Recommendations for estimating the dryness of saturated steam.
BS 3958:(parts)	Thermal insulating materials.
BS 4250:1975	Commercial butane and propane.
BS 5300:1976	Method for the determination of pumpability of industrial oil fuels. (Identical with IP 220/67.)
BS 5410:(parts)	Code of practice for oil firing. (Supersedes CP 3002.)
BS 5422:1977	The use of thermal insulating materials (–40°C to +650°C) (Supersedes BS 1334, BS 1588 and BS 3708.)

CP 331:(parts)	Installation of pipes and meters for town gas.
CP 3005:1969	Thermal insulation of pipework and equipment (in the temperature range of –100°F to +1500°F)
CP 3009:1970	Thermally insulated underground piping systems.

INDEX